QUEBEC HYDROPOLITICS

STUDIES ON THE HISTORY OF QUEBEC/
ÉTUDES D'HISTOIRE DU QUÉBEC
Magda Fahrni and Jarrett Rudy
Series Editors/Directeurs de la collection

1 Habitants and Merchants in
Seventeenth-Century Montreal
Louise Dechêne

2 Crofters and Habitants
Settler Society, Economy, and
Culture in a Quebec Township,
1848–1881
J.I. Little

3 The Christie Seigneuries
Estate Management and
Settlement in the Upper
Richelieu Valley, 1760–1859
Francoise Noel

4 La Prairie en Nouvelle-France,
1647–1760
Louis Lavallée

5 The Politics of Codification
The Lower Canadian Civil Code
of 1866
Brian Young

6 Arvida au Saguenay
Naissance d'une ville industrielle
José E. Igartua

7 State and Society in Transition
The Politics of Institutional
Reform in the Eastern
Townships, 1838–1852
J.I. Little

8 Vingt ans après *Habitants
et marchands,* Lectures de
l'histoire des xviie et xviiie
siècles canadiens / *Habitants et
marchands,* Twenty Years Later
Reading the History of
Seventeenth- and Eighteenth-
Century Canada
*Edited by Sylvie Dépatie, Catherine
Desbarats, Danielle Gauvreau,
Mario Lalancette, Thomas Wien*

9 Les récoltes des forêts publiques
au Québec et en Ontario,
1840–1900
Guy Gaudreau

10 Carabins ou activistes?
L'idéalisme et la radicalisation
de la pensée étudiante à
l'Université de Montréal au
temps du duplessisme
Nicole Neatby

11 Families in Transition
Industry and Population
in Nineteenth-Century
Saint-Hyacinthe
Peter Gossage

12 The Metamorphoses of
Landscape and Community in
Early Quebec
Colin M. Coates

13 Amassing Power
J.B. Duke and the Saguenay
River, 1897–1927
David Perera Massell

14 Making Public Pasts
The Contested Terrain of
Montreal's Public Memories,
1891–1930
Alan Gordon

15 A Meeting of the People
School Boards and Protestant
Communities in Quebec,
1801–1998
*Roderick MacLeod and Mary Anne
Poutanen*

16 A History for the Future
Rewriting Memory and Identity
in Quebec
Jocelyn Létourneau

17 C'était du spectacle !
    L'histoire des artistes
    transsexuelles à Montréal,
    1955-1985
    Viviane Namaste

18 The Freedom to Smoke
    Tobacco Consumption and
    Identity
    Jarrett Rudy

19 Vie et mort du couple en
    Nouvelle-France
    Québec et Louisbourg au
    XVIIIe siècle
    Josette Brun

20 Fous, prodigues, et ivrognes
    Familles et déviance à Montréal
    au XIXe Siècle
    Thierry Nootens

21 Done with Slavery
    The Black Fact in Montreal,
    1760–1840
    Frank Mackey

22 Le concept de liberté au Canada
    à l'époque des Révolutions
    atlantiques, 1776–1838
    Michel Ducharme

23 The Empire Within
    Postcolonial Thought and
    Political Activism in Sixties
    Montreal
    Sean Mills

24 Quebec Hydropolitics
    The Peribonka Concessions
    of the Second World War
    David Massell

Quebec Hydropolitics

The Peribonka Concessions
of the Second World War

DAVID MASSELL

McGill-Queen's University Press

Montreal & Kingston · London · Ithaca

ISBN 978-0-7735-3781-1 (cloth)
ISBN 978-0-7735-3782-8 (paper)

Legal deposit first quarter 2011
Bibliothèque nationale du Québec

Printed in Canada on acid-free paper that is 100% ancient forest free
(100% post-consumer recycled), processed chlorine free

This book has been published with the help of a grant from the Canadian
Federation for the Humanities and Social Sciences, through the Aid to Scholarly
Publications Programme, using funds provided by the Social Sciences and
Humanities Research Council of Canada. Funding has also been received from the
International Council of Canadian Studies through its Publishing Fund.

McGill-Queen's University Press acknowledges the support of the Canada Council
for the Arts for our publishing program. We also acknowledge the financial support
of the Government of Canada through the Canada Book Fund for our publishing
activities.

LIBRARY AND ARCHIVES CANADA CATALOGUING IN PUBLICATION

Massell, David Perera
    Quebec hydropolitics : the Peribonka concessions of the Second World War /
David Massell.

(Studies on the history of Quebec ; 24)
Includes bibliographical references and index.
ISBN 978-0-7735-3781-1 (bound).–ISBN 978-0-7735-3782-8 (pbk.)

    1. Water resources development–Political aspects–Québec–Péribonka
River Region–History. 2. Concessions–Québec–Péribonka River Region–
History. 3. Industrialization–Québec–Péribonka River Region–History.
4. World War, 1939–1945–Québec–Péribonka River Region. 5. Montagnais
Indians – Québec–Péribonka River Region. 6. Péribonka River Region
(Québec)–History. 7. Québec (Province) –Politics and government–1936–1960.
I. Title. II. Series: Studies on the history of Quebec ; 24

TK1427.Q8M386 2010   333.91'4150971414   C2010-904833-4

This book was typeset by Em Dash Design in 10/13.5 ITC New Baskerville

Contents

Maps and Illustrations ix
Acknowledgments xi

Introduction 3

CHAPTER 1
Hydraulic Hinterland,
Native Home: A Prehistory 15

CHAPTER 2
Lac Manouan 35

CHAPTER 3
Passe Dangereuse 77

CHAPTER 4
Beyond Quebec:
the Shipshaw-Massena Bargain 99

CHAPTER 5
"Do we live in the province of Aluminum
or in the Province of Quebec?" 133

CHAPTER 6
Bersimis and Beyond 142

CHAPTER 7
Conclusion 154

Notes 173
Bibliography 219
Index 231

Maps and Illustrations

THE MAPS

Innu family hunting territories for the Pointe-Bleue (Mashteuiatsh)
and Bersimis (Pessamit) reserves, ca. 1920s–30s /xiv
Hydroelectric development, 1890s to 2010 /xv
Saguenay Valley under development /65
The Quebec state looks upstream /66–67
Canada on the US map of war production /68
The war prompts hydroelectric expansion /69
Hydroelectric installations of the Saguenay system /70

THE DAMS

Site of the Lac Manouan storage dam prior to construction /71
Lac Manouan construction by airlift /71
Burning operations, August 1940 /72
Timber cribs under construction, October 1940 /72
Lac Manouan project layout /73
Lac Manouan dam completed /74
Passe Dangereuse access road /74
Passe Dangereuse project layout /75
Passe Dangereuse under construction /75
Completed Passe Dangereuse dam, 1943 /76
Aluminum ingot at Arvida Works /76

THE PLAYERS

C.D. Howe and Herbert Symington, New York, 1941 / 118
J.B. Duke and associates, 1915 / 119
Ray Powell / 120
Ray Powell with Canada's governor general and the bishop of
    Chicoutimi, 1942 / 120
McNeely Dubose / 121
Portrait of Maurice Duplessis, 1936 / 122
Adélard Godbout, 1939 / 122
Godbout and Churchill, 1943 / 123
Portrait of Raymond Latreille / 124
Commissioners of Hydro-Québec taking possession of Montreal Light,
    Heat and Power head offices, April 1944 / 124
Latreille presents Manic 5 at press conference, 1960 / 125

NATIVE PEOPLE

Pointe-Bleue, 1907 / 126
Joseph Dominique and family, 1945 / 126
Bersimis, ca. 1920s / 127
Poling upstream, ca. 1900 / 127
Bersimis hunters pole upriver, late 1930s / 128
Sylvestre Kapu portaging / 128
Sylvestre Kapu and family / 129
Departing by car from Pointe-Bleue / 130
Family of Simon Raphael, ca. 1936 / 130
Fox pelt negotiations, 1930s / 131
Joseph Desterres / 131
Madame Joseph Vachon / 132
Jack Germain, 1979 / 132

Acknowledgments

Sources for this study were hard to come by. All the more reason to acknowledge the valuable assistance of civil servants and archivists who helped me locate relevant material. I thank waterpower specialist Robert Gagnon of the Ministère des ressources naturelles et de la Faune, Charlesbourg, who identified the appropriate historical files of that Ministry. Serge Hamel, Claude Huron, and Claudelle Gauthier helped me to search for documents in the Centre d'expertise hydrique du Québec, Quebec City. Robert Galipeau, Luci Dion, and Nicole Hébert of the Business Information Centre/Centre d'information d'affaires, Alcan Aluminium Limitée (Rio Tinto Alcan), Montreal, showed me copies of several corporate memoirs as well as photographs of the Lac Manouan and Passe Dangereuse dams. At Québec/Sainte-Foy, archivists Pierre-Louis Lapointe, Rénald Lessard, and Nancy Bélanger, and technicien en information Michel Simard skillfully tracked down documents at the Bibliothèque et Archives nationales du Québec; as did archivists Martine Rodrigue and Caroline Rouleau at the Archives d'Hydro-Québec in Montreal, and Estelle Brisson of the Bibliothèque et Archives nationales du Québec in Montreal. Annik Cassista of the Commission de la capitale nationale du Québec helped gather photographs, as did reference technician Louise Caron of Library and Archives Canada. Further afield, Montagnais chief Gilbert Dominique, along with councillors Janine Tremblay and Nelson Robertson, graciously opened their archival holdings at Mashteuiatsh (Pointe-Bleue), while archivist Louise Siméon, researcher Paul Benjamin, and museum director Bibiane Courtois of the Musée amérindien de Mashteuiatsh helped me to navigate this material; and hunter-trapper Gérard Siméon graciously shared his recollections of pre-war life in the bush, as did hunter-trapper Adélard Bellefleur at Pessamit (Bersimis/Betsiamites). At the Pessamit Reserve, I am deeply grateful to Jean-Marie ('Jack') Picard of the Conseil des

Innus de Pessamit for opening the collection of the Council's Secteur negotiation, and to Antoine Bacon and Pascal Richard for helping me to locate relevant interviews, reports, and maps among its contents. The warm hospitality of Jack and his wife Thérèse Rock-Picard, along with their personal knowledge of Innu history and reserve life, made my research trip to Pessamit both rich and memorable.

Various historians and other professionals helped me to craft a manuscript that accurately reflects the research. The Saguenay's pre-eminent historian, Camil Girard, helped me to better understand the history of the region's Native peoples, as did anthropologists Denis Brassard and Paul Charest. Quebec's seasoned economic historian Claude Bellavance kindly read and wisely commented on an earlier version of this manuscript. John Herd Thompson and José Igartua, both top-flight historians of Canada, shared their insights at several troublesome points during the course of this research. I am grateful to Kitty Samuel, Jim Shaw, Sandy Savoca, Mike Caufield, and Mike Hawes of Alcoa's Massena Operations, as well as Theresa Sharp of the Massena Museum, for help in piecing together Massena's industrial history. Two McGill-Queen's University Press anonymous readers provided constructive critiques of methodology, interpretation, and style, based on close readings of the manuscript, as did Jarrett Rudy and Magda Fahrni, editors of the McGill-Queen's University Press series Studies on the History of Quebec. McGill-Queen's publication team, including Joan McGilvray, Joanne Pisano, Jonathan Crago, Elena Goranescu, and Ryan Van Huijstee, as well as copy editor Eleanor Gasparik, expertly shepherded this work from manuscript stage to book. It was a pleasure to work with mapmaker Mike Hermann of Purple Lizard Maps; Mike and I also benefited from the advice of geographer Étienne Govare of Hydro-Québec Équipement, Géomatique division. Historians and geographers Matthew Evenden, Scott See, Jean Martin, Tina Loo, Mark Fiege, Hans Carlson, Jim Kenny, and Bill Parenteau offered useful critical advice. I am grateful to the members of the Département de géographie of Université Laval as well as the Centre interuniversitaire d'études québécoises (CIEQ), especially historical geographer Matthew Hatvany, for an invitation to spend a sabbatical year conducting research at/around Laval University. I am equally grateful to my colleagues in the Department of History at the University of Vermont (UVM): especially, Flora Cassen, Sean Field, Frank Zelko, Jon Huener, Sean Stillwell, Amani Whitfield, and Paul Deslandes. Richard and Colin Gliech of Burlington, Vermont, provided translation help, as did my colleague in the American Council for Québec Studies,

Amy Reid. So did friends Hélène Laprise and Laurian Brunet of Quebec City, who also generously provided lodging and hospitality during several research trips toward the end of this project. At UVM's Canadian Studies Program, I am grateful for both the financial and intellectual support of Bill Metcalfe, André Senécal, and Paul Martin.

Conducting and publishing one's research requires both financial and moral support. The Government of Canada provided generous monetary support, in the form of a Canadian Studies Faculty Research Grant, as did the Government of Quebec, in the form of two of its Grants for Professors and Researchers. UVM's Dean of the College of Arts and Sciences offered research support through the Dean's Fund for Faculty Development. This book has been published with the help of a grant from the Canadian Federation for the Humanities and Social Sciences, through the Aid to Scholarly Publications Program, using funds provided by the Social Sciences and Humanities Research Council of Canada. Publication was also made possible by a grant from the International Council for Canadian Studies (ICCS) through its Publishing Fund. Portions of chapters one and four appeared previously in *Québec Studies* and *The American Review of Canadian Studies*. I thank the editors of both journals for permitting me to reshape that material here. Last and deepest thanks is due to northern paddlers and friends Michael Secor, J. Ladd, and Martha Whitney; my sister, Deborah, and her partner, Jackson Francisco; Sylvia Brinton Perera, my mother and an original thinker and scholar; my father, Gregory Massell, who, despite a recent stroke remains a sharp-eyed editor; my children, Emily, Abby and Kate, who built the foundations of a second language by spending a year in Quebec during the research phase of this project; and my sage and spirited wife and partner, Laura Nault Massell.

xiv

Innu Family Hunting Territories for the Pointe-Bleue and Bersimis Reserves circa 1920s-30s

HERVIEUX · TSHERNISH · JOURDAIN · PICARD · PICOUTLAIGAN · SAINT-ONGE · VACHON · COPEAU · VACHON · JOURDAIN · Lac Pletipi · CREPEAU · COLLARD · Eastmain · Savane · WASHISH · BOIVIN · BACON · DESTERRES · SAINT-ONGE · SAINT-ONGE · Michel · L'ABBE · ANDRE · DESTERRES · COLLARD · Montagne Blanche · BENJAMIN · HERVIEUX · Aux Outardes · TSERNISH · BENJAMIN · NATIPI · FONTAINE · SAINT-ONGE · Modeste · Grande Loutre · Péribonka · ROCK · Lac Manouane · BACON · VACHON · Témiscami · Rupert · DOMINIQUE · PICARD · Manicouagan · PICARD · Lac Albanel · Lac Onistagane · TSHEKETASH · SIMON · St-Onge · BACON · ASHINI · BOSSUM · RIVERIN · DESTERRES · Lac Mistassini · MACDONALD · ÉTIENNE · HERVIEUX · PICARD · SYLVESTRE · Mistassibi · Serpent · RÉGIS · ROCK · VOLLANT · JOURDAIN · Manouane · VALLEE · HERVIEUX · McKENZIE · Broadeuse · BELLEFLEUR · RIVERIN · BACON · JOURDAIN · Mistassini · XAVIER · FONTAINE · RIVERIN · HERVIEUX · Baie Comeau · GERMAIN · Lac Pipmuacan · Betsiamites · SAINT-ONGE · CANAPE · ROUSSELOT · VALLEE · MATABEE · JOURDAIN · SIMEÓN · THOMAS · VOLLANT · Pessamit (Bersimis) · MANIGOUCHE · VERREAULT · CANAPE · PAUL · Sault aux Cochons · Ashuapmushuan · NAPANEE · CONNOLLY · Shipshaw · HERVIEUX · RAPHAËL · XAVIER · Portneuf · St. Lawrence · Mistassini · Dolbeau · BEGIN · Honfleur · Lac Saguenay · SIMÉON · VERREAULT · Saint-Félicien · Lac St. Jean · Alma · Chicoutimi · Bagotville (La Baie) · Mashteuiatsh (Pointe-Bleue) · Roberval · Jonquière · BOIVIN · NACOUTI · Quebec · CANADA · UNITED STATES

▲ Native Community
◦ Population less than 4999
◉ Population 5000–9,999
◉ Population 10,000–99,000
✪ Capital City, population 100,000+

1931 Population Data

0 — 100 Miles
100 Kilometers

D. Massell, M. Hermann ©2010

The Saguenay and North Shore watersheds were home to numerous Innu/ Montagnais families until their displacement in the decades after the Second World War. The location of hunting territories represents the interwar era. Watercourses are shown in their original state, prior to hydroelectric development.(© D. Massell, M. Hermann, 2010, information drawn from Brassard, *Occupation et utilisation du territoire*, 1983; Deschênes and Dominique, *Nitasinan*, 1983; and Frenette, *Occupation et utilisation du territoire*, 1983)

Hydroelectric Development, 1890s to 2010

Petit Lac Manicouagan Reservoir 1960
Hart-Jaune 1960
Lac Manicouagan Reservoir 1971
Ste-Marguerite 3 Reservoir 2000
Lac Pletipi
Manic 5 1968
Manic 3 Reservoir 1976
Ste-Anne Reservoir 1958
Reservoir Toulnustouc 2005
Bonnard 1961
Lac Manouane Reservoir 1941
Lac Onistagane
Passe Dangereuse Reservoir (Lac Péribonka) 1943
Chute-des-Passes 1960
Péribonka 4 2008
Outardes 4 Reservoir 1969
Outardes 3 1969
Bersimis 1
Ripmuacan Reservoir 1956
Bersimis 2
Manic 2 1965
McCormick 1950-51
Manic 1 1966
Baie Comeau
Outardes 2 1978
Bersimis (Pessamit)
Eastmain
Rupert
Lac Albanel
Lac Mistassini
Ashuapmushuan
Témiscamie
Péribonka
Mistassini
Manouane
Aux Outardes
Manicouagan
Toulnustouc
Ste-Marguerite
Betsiamites
1956 / 1959
Sault aux Cochons
Portneuf
Shipshaw
Chute-du-Diable 1953
Shipshaw 1943
Chute-à-la-Savane 1953
Saint-Félicien
Dolbeau
Mashteuiatsh (Pointe-Bleue)
Lac St. Jean Reservoir 1926
Kenogami Reservoir 1924
Alma
Honfleur
Roberval
Jonquière
Chicoutimi
Bagotville (La Baie)
Saguenay Fjord
St. Lawrence
Quebec
CANADA / UNITED STATES

Legend:
▲ Native Community
◦ Population less than 4999
◉ Population 5000–9,999
◉ Population 10,000–99,999
✪ Capital City, population 100,000+
▭ Power Plant
◼ Storage Dam → Diversion
◢ Storage Dam and Power Plant
🖌 Reservoir
2001 Population Data
0 — 100 Miles
100 Kilometers
D. Massell, M. Hermann ©2010

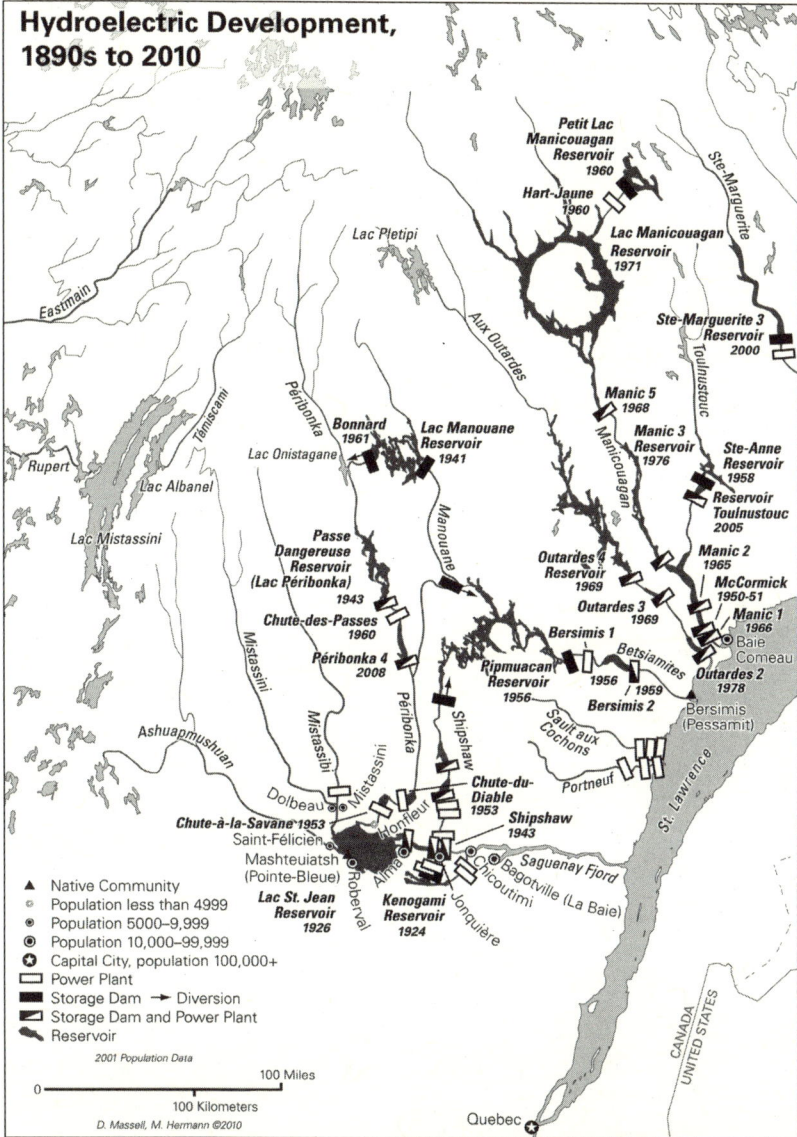

The construction of storage dams at Lac Manouan and Passe Dangereuse inaugurated a profusion of post-war hydroelectric projects in the backcountry. (Names and dates are provided for the largest projects. Tributary streams have been removed to show dam/reservoir locations more clearly.) Such projects opened the interior to forestry, mining, and sport hunting. They also rendered river valleys unfit for Native hunting and travel, and catalyzed social change among the Innu, drawing Aboriginals off the land to wage labour and sedentary reserve life. (© D. Massell, M. Hermann, 2010, information drawn in part from the map "Installations et territoires de production," Hydro-Québec, 2004)

QUEBEC HYDROPOLITICS

Introduction

In late summer of 1938, twenty-five-year-old Montagnais/Innu[1] trapper Louis-Georges Boivin set out with his brothers from the Indian reserve at Pointe-Bleue, Lac St-Jean, on an 800-mile journey by canoe and snowshoe. The young men had spent several weeks at the reserve with family and friends, sharing stories of the previous season's hunt, attending Mass and exchanging pelts for a trapper's provisions at the trading post. They would spend the next eleven months, through autumn, winter, and spring, roaming the northern bush, tending their traplines and hunting game. To the well-fed, urban-dwelling historian of the early twenty-first century, the Boivins' voyage might be regarded as 'epic' in its hardship and duration. To the Boivins, it was a long-established and economically essential annual migration of a semi-nomadic people still very much engaged in the seasonal rhythm of the Canadian fur trade.

The Boivins began their journey by truck, rounding the shore of Lac St-Jean, Quebec, to the Peribonka River where they proceeded by wood-canvas canoe. Within several days, they left behind farm country and logging operations, and entered the boreal forest, paddling, poling and portaging northward on the lake's largest tributary, the Peribonka (Innu for 'it flows through sand'). Within the week, at the confluence of the Peribonka and Manouan rivers, the brothers split up: two continued up the Peribonka, while Louis-Georges and brother René moved eastward up the Manouan ('place where we gather eggs'). Each of these two young men paddled a canoe, and each canoe was loaded with half a ton of provisions that included flour, lard, tobacco, tea, and ammunition. Despite the load, the Boivin brothers made rapid progress. In less than two months' hard travelling via the Manouan and White Mountain rivers, they reached the height of land of the Rivière aux Outardes. To this point the brothers lived largely off the *petite chasse* [small game] of hare, partridge, duck, and fish. Once at their hunting ground, they also

shot moose, where available, and tended fishnets beneath the ice. The
men constructed a log cabin, cut firewood, and fashioned snowshoes
and toboggans from local birch trees. As winter arrived, they departed
the main camp for weeks at a time to walk the neighbouring hills and
creek valleys of their territory, sleeping each night in a canvas tent. The
two Natives harvested mink, marten, fox, wolf, and beaver. Following
the holiday *fêtes*, they roamed further afield into the headwaters of the
Manicouagan River, trapping marten, mink, and lynx. With the breakup
of winter's ice, the young men continued to trap beaver and otter by
canoe, and to hunt ducks and geese. At season's end, they loaded fur
bales into the canoes and cached everything else, including tents, stoves,
traps, and winter clothes. The Boivins then cautiously descended the
swollen Manouan and Peribonka rivers back toward the trading post
and the reserve.

From September through June, the Boivins had encountered several
other Aboriginal families from Pointe-Bleue, and at least one family
from the North Shore's Bersimis Reserve, but they had seen no signs of
'*l'homme blanc*' [the white man] Although this year-long journey might
seem an eighteenth- or nineteenth-century recounting, it happened
almost mid-twentieth century. At this historical moment, on the eve of
the Second World War, Quebec's northern interior remained a thoroughly Aboriginal world.[2] However, unbeknownst to the Native Peoples,
change was coming, change directed by both government and corporate
interests.

The provincial government of Quebec had laid the groundwork
for hydroelectric development in this watershed from the turn of the
twentieth century. Lac St-Jean itself was no longer a natural lake. It
served as a storage reservoir, impounded and enlarged by American
businessmen in the 1920s, with government approval, to make large
quantities of hydroelectricity downstream on the Saguenay River. The
project's chief engineer touted this Isle Maligne power plant (completed
in 1926) as the "largest single installation in water-power development
ever undertaken." And the dam's huge power-generating capacity was
due in good part to the water stored upstream; impoundment of Lac
St-Jean more than doubled what engineers call "the firm, regulated discharge" of the Saguenay River throughout the year, from some 12,000
cubic feet per second of flow to 30,000 cubic feet per second. The same
Americans had dammed the lake without informing the local French-
Canadian farmers or compensating them when their lands were flooded.
The resulting outcry over '*la tragédie du lac Saint-Jean*' had prompted

attentive provincial government civil servants to consider measures of flood control to protect their rural constituents from further damages. Quebec bureaucrats pondered: why not reduce flood damage to an agricultural region (farmers voted, after all) by building reservoirs further upstream in 'Indian country' (Aboriginals didn't vote)? In this way, the unpleasant consequence of industrial development was passed down the early-twentieth-century social hierarchy from American industrialist to Quebec citizen to nomadic Native. Through the early 1930s, provincial land surveyors mapped the coordinates of latitude and longitude northward into the Peribonka valley to bring this great hydraulic basin within the purview of the state, rendering it "legible"[3] for both bureaucrats and businessmen. Quebec's hydro specialists identified the most promising locations for dam sites and reservoirs that would better regulate the flow regime of the entire Saguenay watershed and facilitate its continued industrial development.

The prospect of war would sharpen corporate interests in actually building the dams. By the onset of the Montagnais's 1938 hunting season, Japan had launched a full-scale invasion of China and completed its "rape of Nanjing," an early indication of the huge wartime bloodletting to come. As Louis-Georges Boivin and family ascended the Peribonka River that autumn, British prime minister Neville Chamberlain met with Adolf Hitler in Munich and proclaimed "peace for our time." Germany, far from appeased in its imperial aspirations, invaded the Sudetenland and then the rest of Czechoslovakia. As the Boivins built their cabin of black spruce logs at the height of the White Mountain River, Nazi storm troopers and Hitler youth vandalized and terrorized Germany's Jews in the infamous Kristallnacht ('night of broken glass'). And as the Native hunters tended their traplines in careful circuits through the late winter months, Hitler directed the Army High Command to ready an attack on Poland for September. By the start of the trapping season of 1939, Europe plunged into war.

Industrial armies waging war require strategic materials. The light, flexible metal aluminum, essential for modern aircraft, proved a key ingredient for both Axis and Allied forces. Aluminum Company of America spawned Aluminium Limited (later, with Aluminum Company of Canada, to become Alcan) in 1928 to assume control of international production based in a dominion of the British Empire. While Aluminum Company of Canada executives had seen ingot orders evaporate during the Great Depression, rearmament reversed this trend beginning in 1936, with the organization happily making deliveries to Japan and

Germany as well as Britain and the United States. Aluminum ingot production requires large quantities of cheap power, usually from hydro-electricity. Hydroelectric energy, in turn, is derived from large volumes of falling water. Therefore, as early as 1937, Aluminum Company of Canada engineers were scouting dam sites for reservoirs high in the Peribonka basin; by late 1938, engineer-executives were pressing the government of Quebec for the concession of the necessary Crown lands for these water storage facilities.

Large resource projects, and perhaps hydroelectric dams in particular, form a marvelous prism through which the historian may examine the past. They reveal, as though in refractive layers, the several constituencies or pressure groups that demanded or contested major landscape changes in a given era as well as the dynamic interaction between them. The Boivin brothers, and hundreds like them, sought a living from the northern forest in its current state, as had those before them for roughly 5,000 years. Dubbed *Montagnais* ('the mountain people') by the French, they called themselves *Innu* ('the people'). Meanwhile, a democratic-capitalist state sought to reorder, to reorganize, what it deemed 'natural resources' for the financial and political benefit of its voting constituents. Toward this end, Canadian and Quebec functionaries and politicians appear in this study as managers and regulators of the exploitation of hydraulic resources, working in the service of provincial and/or national economic development and making possible the vigorous prosecution of global war. Nor are the events of this study limited to Canada. After 1940, Canadian war planners increasingly coordinated industrial production with their counterparts in the United States. Therefore, our research ranged further afield in order to tease out American wartime demand for strategic materials, as well as the means and mechanisms by which Americans and Canadians sought to mobilize such resources. As we will see, the actions and decisions taken in Washington, DC, would leverage decisions that were made in both Ottawa and Quebec City. Equally important in this study is an influential corporate entity of American origin, which was responding to surging demand for war materiel in Europe and beyond. Aluminum Company of Canada, the dam builder, already held rights to a portion of the Saguenay watershed's hydroelectric energy and now sought to expand its jurisdictional domain northward into the hinterland.

In integrating these multiple perspectives on landscape change, our narrative cannot fall neatly into one of the historical subdisciplines such as business history, political economy, or social history. It is instead, and by necessity, a blend of all three.[4] While the reader may find this approach somewhat novel or innovative, it is not our intention to demonstrate methodological novelty per se. Our aim has been to reconstruct a story that enjoined human actors from widely different North American milieus and with varied motivations. Washington's war planners looked north to Canadian resources to meet escalating demand for metal and power; Ottawa's politicos, increasingly interventionist in provincial affairs during wartime, sought to fulfill these demands by tapping the hydraulic wealth of the province of Quebec; Quebec's hydro bureaucrats and politicians set out to ensure that the dams built by an American subsidiary corporation would compensate the needs of Quebec citizens, in particular the French-Canadian majority; and as the Peribonka concessions energized Quebec's debate over electricity's nationalization, nationalist civil servants pushed provincial politicians to fashion an institutional instrument of French-Canadian liberation in Hydro-Québec. Other Quebec citizens, not yet accounted for in the democratic process, tended traplines along river valleys that were soon to be flooded.

Quebec's history is entangled in, and quite inseparable from, a wider North American history in the pages that follow. Quebec's 'national' story – the French-Canadian struggle to assert control over economic affairs, in particular hydraulic development, as a means of ethnic self-determination – is here entwined with and very much coloured by alternative narratives, ranging from US continental investment to northern Aboriginal life.

The particular setting for this human and institutional contest is a northern watershed. The result of dam building was profound and permanent change to a riverine environment, in addition to the social world that it sustained. And so, at the geographic centre of our study are two broad puddles on Earth's surface, water storage reservoirs impounded in the Peribonka valley to feed hydroelectric power plants downstream along the Saguenay. This is a study of a landscape's origins. Our purpose is to inquire into the negotiating process by which these artificial lakes were conceded by the state to private industry in the midst of the Second World War. As such, this book is a sequel of sorts. Of the three expansive man-made lakes that regulate the flow of Quebec's great Saguenay River for the purposes of power generation and industrial manufacture, only Lac St-Jean has drawn the attention of scholars. My own work, *Amassing Power: J.B. Duke and the Saguenay River, 1897–1927*, carefully documents the process by which the Quebec government and its bureaucrats condeded this reservoir's creation and use to private industry.[5] *Quebec Hydropolitics* simply moves upstream and also forward in time to reconstruct the concession of the remaining two, lakes Manouan and Passe Dangereuse, located in the upper reaches of the Saguenay's principal feeder, the Peribonka River.[6]

The Peribonka River rises on the granite and muskeg of the Canadian Shield some 300 sinuous miles north of Lac St-Jean, and eventually spills its waters into the lake itself. Comprising about one-third of the drainage area of the Saguenay basin as a whole (the entire basin is roughly the size and shape of the state of Maine and the Peribonka drains 12,000 of its some 30,000 square miles), its regulation is clearly an important factor in the control of the watershed. Readers may be familiar with the name *Peribonka*, as the lower valley was the setting for French author Louis Hémon's celebrated novel of 1914, *Maria Chapdaleine*. Our concern in this study is not with the Peribonka's short band of arable land adjacent to Lac St-Jean (about 20 miles of the river's 300), but rather its vast boreal and subarctic uplands. And our focus is not on agricultural colonization of the late nineteenth and early twentieth centuries, as was Hémon's, but on the process that was fast supplanting it in Quebec: industrialization.

Why chronicle the industrial history of a remote Quebec river? To begin with, flooding large tracts of the northern forest, as well as altering the natural flow of the river, held human-environmental consequences, touching the lives of the local Montagnais trappers[7] as well as the interests of forestry and fisheries. Further downstream, the Peribonka's damming and regulation boosted the power capacities of the Saguenay's generating plants. By current technological standards, lakes Manouan and Passe Dangereuse are modest in scale; neither bulks as large on the map of Eastern Canada as, say, the hydroelectric reservoirs of the Manicouagan (1960s–70s), Churchill Falls (1970s), or James Bay (1970s–present) projects. Neverless, both were formidable projects in their day, with Passe Dangereuse then being the largest storage reservoir by volume in the hydro-rich province of Quebec. Together, these water-storage bodies would raise the Saguenay's discharge to some 42,700 cubic feet per second, once again roughly doubling the power-generating capacity of the Saguenay's hydroelectric plants downstream. The *Canadian Geographical Journal* celebrated the result: by the end of the Second World War, with some 2,000,000 horsepower of electricity online, Quebec's Saguenay Valley had become "the greatest aluminum centre in the world."[8] Additional water storage greatly enlarged not only

Aluminum Company of Canada's production of ingot for a continental arsenal in aluminum during the Second World War,[9] but also the company's assets and territorial domain. "The war transformed a marginal aluminum producer into a giant," mineral economist Carmine Nappi has written.[10] With the Peribonka reservoirs secured in long-term leases from the Quebec government in the early 1940s, Aluminum Company of Canada consolidated its control over the hydraulic resources of the large part of the Saguenay watershed. That assured the company a regional monopoly over the basin's hydraulic resources and a source of cheap, renewable energy, which helped to make it a major player in the world's aluminum industry in the post-war period. (In 2007, British-Australian multinational mining giant Rio Tinto acquired the now-named Alcan Aluminium Limited for $38 billion, creating Rio Tinto Alcan, "the new global leader in aluminium." Its headquarters remain in Montreal.[11]) And, of course, the Peribonka's development, in concert with that of the Saguenay, brought large-scale industrialization in the form of aluminum and newsprint production to what was, until the 1920s, still a decidedly rural region of Quebec. As farms sprouted factories, and farmers became labourers and consumers, an agricultural outland was transformed into an increasingly industrial and urbanized centre.[12]

Documenting the Peribonka concessions can serve to enrich our understanding of the history of an important Quebec region; more broadly, this study speaks to the political economy of hydroelectricity in the province as a whole. The historiography of the hydroelectric industry in Quebec has been described in detail elsewhere.[13] It is sufficient to note here that the vast majority of a growing literature focuses on events of the post-war era: whether on the institution of Hydro-Québec itself post-1960s,[14] or on that institution's most grand (and controversial) achievement of the following decade, the James Bay Hydroelectric Project.[15] Far less scholarship has dwelt on the pre-war period;[16] less, still, explores the nature of hydro concessions to private enterprise in this formative era, as well as the origins of administrative regulation that gave rise to the creation of Hydro-Québec;[17] and the period comprising the Great Depression and the Second World War remains an all but empty slate.[18] Negotiations regarding the Peribonka River spanned precisely these periods, which take in Maurice Duplessis's first mandate and the administration of Adélard Godbout. It was in 1940 and 1943, respectively, that the government of Quebec granted/leased to Aluminum Company of Canada lakes Manouan and Passe Dangereuse. Measured by the resulting changes to the natural landscape and by the political fallout from the arrangements, these grants marked the most significant natural-resource decisions of the wartime period. This case study, focusing on the government's Hydraulic Service within the Department [or Ministry] of Lands and Forests, may serve to advance and also clarify our knowledge of the provincial hydro policy and politics that produced Hydro-Québec.

The timing of these decisions should also be of interest to students of Quebec history. Paradoxically, even as the Quebec state moved to assert its own role as a producer and distributor of electrical power in the formation of Hydro-Québec, it permitted private enterprise, on the eve of this historic decision, to retain and very much enlarge its control of one of the province's most energy-productive basins. This, in effect, marked the bifurcation of Quebec's precious hydraulic resources into private versus public domains. It is important to note that this drama involving public and private power was played out in wartime. Lac Manouan and Passe Dangereuse were conceded and created to serve wartime armament contracts, and these concessions were made during a period of rising nationalist demand for public ownership and development of waterpowers as a means of achieving French-Canadian economic 'liberation.' Hence, the most important questions of this study: Who participated in these decisions of the 1940s to refashion Quebec's hydraulic domain, and who did not? How did the war's sense of emergency shape the application of resource policy in Quebec in this era? To what extent did the war accelerate hydrographic change in the case of the Saguenay basin? What was the relationship, if any, between the Peribonka concessions and the creation of Hydro-Québec? In what ways might this discrete study of a single watershed speak to the history of Canadian northern development, and/or the impact of war on Canadian Aboriginals, or the interaction of war and the natural environment, an emerging subfield of environmental history?

While the geographic centre of this study is the boreal/subarctic landscape, at its methodological heart is a rich series of provincial government files held in the working offices of Quebec's Ministry of Natural Resources and Wildlife (Ministère des Ressources naturelles et de la Faune) in Charlesbourg, Quebec. For a full half-century before the formation of Hydro-Québec, the government of Quebec relied on private entrepreneurs to make useful development of the Crown's hydraulic resources. At first, the state sold waterfalls outright; after 1910, bowing to nationalist criticism, it began parting with the province's riparian rights under the negotiated terms of an emphyteutic lease (a long-term contract requiring the lessee to develop or improve the property in order to increase its value). Such leases required that the state's newly formed bureaucracy, the Hydraulic Service, maintain in orderly fashion its contracts with private firms along with all and sundry correspondence and memoranda in this regard. Thus, the Service kept careful record of each Quebec watershed and its lessees, whether Aluminum Company of Canada at the Saguenay, Shawinigan Water and Power at the St-Maurice, Gatineau Power or MacLaren-Quebec Power at the Ottawa. First identified and used in the course of my doctoral research at Duke University in the 1990s in a study of the Saguenay River and Lac St-Jean, this material has continued to be invaluable as historical documentation of Quebec's hydraulic-industrial history.

These files have been augmented by a variety of other sources. Most important: relevant material passed down from this Ministry to the Bibliothèque et Archives nationales du Québec, Sainte-Foy/Quebec City branch; available materials at Alcan Aluminium Limited (Rio Tinto Alcan), in Montreal; the papers of former Quebec premier Adélard Godbout; Quebec newspapers and government publications; the reconstituted debates of Quebec's Legislative Assembly, including those available online at the Bibliothèque de l'Assemblée nationale; various collections in Library and Archives Canada, the Franklin D. Roosevelt Presidential Library in Hyde Park, New York, and the National Archives and Records Administration of the United States in College Park, Maryland; a mass of documentation at the Musée amérindien de Mashteuiatsh on the shores of Lac St-Jean and, on Quebec's North Shore, at the Native reserve/village of Pessamit (Betsiamites); Hudson's Bay Company Archives; and the scholarship dealing with the history of the Saguenay-Lac-St-Jean region, with Canada's North generally, with natural resource policy and practice under the Duplessis and Godbout administrations, and with a small but growing body of thought about the relationship between war and environment.

The Native sources merit further explanation. In the late 1970s, with the era of environmentalism and Native claims under way, Quebec's Montagnais (Innu), in collaboration with the Attikamek, set out to document their occupation and use of ancestral territory as part of the legal preparations for land claims negotiations (still pending) with the government of Quebec. Lacking the customary written documentation of western societies, the Natives instead tapped a rich oral tradition by interviewing living hunter-trappers of each community. The gathering of

data and identification of possible informants was authorized and organized by the local Conseil de bande and the Council's Comité de chasse
et de trappe. The interview process was supervised by an anthropologist
who trained college-age Innu students to conduct the interviews and
who would eventually synthesize the mass of data into a summary report.
During the interviews themselves, the young interviewers recorded geographic information (routes and campsites) onto Canadian topographic
maps, and filled in one note card (or *fiche*) for each campsite described
by the hunter (noting the length of time spent at the place and the
animals harvested). All interviews were recorded on cassette tape, and
a selection were transcribed and/or translated into French.[19] It is these
interviews, as recorded on maps, summarized on *fiches* and, in some
cases, fully transcribed, that I have tapped to reconstruct something
of pre-war Native life on the terrain that would be rearranged by the
construction of hydroelectric dams.

    As to the reliability of this source material: could a hunter-trapper in
his sixties or seventies recall, when interviewed in 1981, the routes and
particular events of a journey that took place in the 1930s, and do so
in a way that would clear the historians' bar for accuracy? In this historian's view, the answer is an unequivocal yes: first, because the hunters'
lifework was to travel such routes year after year, and the subsistence of
their families depended on accurate memory of what fish, bird, or mam-
mal was likely to be found in which river eddy or creek valley or along
which portage trail; and second, because the contents of the interviews
were cross-checked, as needed, with others of the same hunting party
to assure a high degree of veracity. Moreover, my own intention was not
to craft a hunter's biography, which might well be problematic based
on these sources alone, or a comprehensive history of the Pointe-Bleue
or Betsiamites reserves. Instead, it has been to identify and document
the evolving patterns of pre-war Native land use in the Manouan and
Peribonka uplands, with an eye toward incorporating this information
into a study of resource politics and policy. In short, humans lived here;
surely we need to understand who they were and how and where they
lived; surely we should explore how and whether these local inhabitants participated in, or in any way shaped, decisions that reconfigured
a hydraulic landscape. Toward this end, a close reading of dozens of
hunters' accounts amply served my purpose. With the use of maps and
fiches alone, it became a straightforward task to reconstitute a hunter's
annual migration; and where transcripts were available, I was able to
nuance the narratives with the words of the hunter him/herself. When

quoting from the transcripts, for the most part I have chosen to translate the original French into English (as I have also done for Quebec governmental correspondence); where translation was a matter of any judgment, I have preserved the French in the notes.

The richness and volume of these Native sources stands in marked contrast to the paucity of material made available by the Saguenay's corporate landlord. The Innu of the Mashteuiatsh and Pessamit communities, amidst ongoing land claim negotiations with Quebec, nonetheless opened their collections to an American historian who was trying to track the origins of landscape change. Aluminum Company of Canada, a major player in this study, was less co-operative. In fact, this book's principle shortcoming, in the author's view, is due to lack of access to the internal correspondence, including board of directors' meeting minutes, of this corporation. Such archival material, if it in fact still exists,[20] may well have allowed me to make a more detailed rendering and analysis of the corporate perspective on wartime hydroelectric expansion in Quebec regarding executives' negotiations for additional concessions at the Saguenay and the firm's relationships with Quebec, Canadian, and American governments during the 1930s and 1940s. Alas! Whether the company's intention was to avoid legal entanglement or simply to keep what may be unsightly corporate behaviour hidden under a rock, the end result was the same. Despite multiple requests over the course of several years to examine this material, I, along with others, was turned away. Rio Tinto Alcan's explanation: the corporate correspondence I sought from the interwar and wartime period constitutes "strategic" and therefore "confidential" material whose contents must be kept "unknown from competition."[21]

A WORD ABOUT CORPORATE HISTORY AND NOMENCLATURE

What grew into the American aluminum giant Alcoa was founded by a young chemist and recent Oberlin College graduate, Charles Martin Hall, in 1888 in Pittsburgh, Pennsylvania, as Pittsburgh Reduction Company. In 1899, now backed by bankers/investors Andrew and Richard Mellon, this US firm established a Canadian beachhead at Shawinigan Falls, Quebec, dubbing its subsidiary Northern Aluminum Company, using the American spelling "aluminum" rather than the Commonwealth and European "aluminium." In 1907, the parent firm changed its name to Aluminum Company of America, soon abbreviated to Alcoa. In 1925, as Alcoa expanded to the Saguenay basin, Northern Aluminum Company was renamed Aluminum Company of Canada, Limited (later abbreviated to Alcan). Three years later, Alcoa placed most of its foreign properties, including those in Canada, into an independent holding company called Aluminium Limited "to engage in the international aluminum business." While the European/Commonwealth spelling of the name reflected the company's intention to do business throughout the world, its directors and stockholders actually remained largely those of Alcoa through the 1940s.

As for the acronym Alcan: in 1944, Alcan became the principal trademark of Aluminium Limited, stamped onto its ingots. It gained increasing recognition on the world's stock exchanges, so in 1966, Aluminium Limited adopted the Alcan trademark as the name for the entire Canadian firm: Alcan Aluminium Limited or, in French, Alcan Aluminium Limitée.[22] With Rio Tinto's acquisition, the name was adjusted yet again. It is interesting to note that Alcoa attempted a hostile takeover of Alcan Aluminium Limited in 2007 that was only resolved when Rio Tinto's friendly acquisition was formalized. Had Alcoa's bid been successful, the one-time Canadian subsidiary would have returned to the parent firm.

In the following pages, Alcan is used to describe Aluminum Company of Canada, Limited, despite the fact that the term was not yet in general use during the years of the Second World War. Furthermore, we will use, quite interchangeably, the terms most frequently used by contemporaries for this corporation: Aluminum Company, the Aluminum Company, or simply Aluminum.

1

Hydraulic Hinterland, Native Home: A Prehistory

This chapter focuses on the Peribonka's industrial 'prehistory,' that is, on the period before the state conceded rights to the river's development and before actual dam building took place. How were human beings utilizing this terrain prior to its industrial transformation? When and why did capital first seek to develop the energy resources of the Peribonka River? How did the state respond to capital's demands? Finding answers to these questions promises to fill out our knowledge of the economic development of the Saguenay region. It will yield insight into the evolving policy and politics of hydro development in the decades before the birth of Hydro-Québec. Finally, this chapter permits us to watch the northward creep of the hydraulic frontier into a territory occupied until then by Native people.

The Peribonka valley remained the most remote quarter of an already-remote North American outland, intensely isolated from Euro-Canadian civilization, well into the twentieth century. It was the very last sector of the Saguenay watershed to see agricultural colonization, which proceeded only after 1887, a full half-century after farming settlement began in the Saguenay region generally. By the time Frenchman Louis Hémon spent the summer of 1912 observing French-Canadian farm life in the valley's lowest reaches, total white population hovered around 1,000 permanent residents. Farms were clustered around several small villages (Saint-Henri-de-Taillon, Péribonka, Sainte-Monique-de-Honfleur) situated at or below the Peribonka's fall line and so able to access steamship travel on Lac St-Jean. Resource extraction in the valley did not proceed much more rapidly than farming. The occasional explorer/surveyor, seeking to inventory timber or minerals on behalf of the Quebec or Canadian state, had viewed only snippets and sections of the river by the late nineteenth century. Such parties included surveyor-geologists John Bignell and A.P. Low of the Canadian Geological Survey, who relied on Native guides to traverse the Peribonka's uplands from the North Shore in 1884 en route to Lake Mistassini. Anglo lumber baron William Price cut timber in the river's lower reaches beginning in the 1850s. His grandson, William Price III, under the corporate name Price Brothers, leased additional timber limits from the Province after 1900 for a growing pulp and paper operation well downstream at Jonquière, as did competing timber merchant Benjamin Scott, who operated a sawmill at Roberval. As late as the 1930s, however, their wood-cutting operations reached no more than fifty miles upstream of the mouth of the river. The Peribonka's vast and wooded boreal hinterlands remained untouched by commercial timber interests right through the Second World War.[1]

As to the Peribonka's hydro resources, it was the First World War before entrepreneurs looked to tap those; and that demand must be understood in the context of businessmen's evolving interests in the Saguenay watershed as a whole. When Canadian entrepreneurs first conceived of harnessing the immense energy of the Saguenay at the turn of the last century, their approach was piecemeal, looking to exploit individual rapids or falls along the river's 30-mile length. When American tobacco magnate James Duke arrived to view the river in 1912, he brought not only the necessary financing to carry out the Saguenay's development but also a shift in the scale of intended development schemes. Duke's hydroelectric engineer was American William States Lee, chief engineer of Duke's Southern Power Company and a man who stood at the pinnacle of his field. In the US South, along the Catawba River, Lee had helped to pioneer the art of comprehensive watershed development for electrical power generation. Upon encountering the Saguenay, he applied his Catawba experience to a gigantic river of the Canadian Shield.

By early 1914, working at his offices in Charlotte, North Carolina, Lee planned an aggregate development of the entire Saguenay in two great stages. The first, which Duke saw to completion before his death, was the Isle Maligne Station located some ten miles below Lac St-Jean and designed to amass the first 110 feet of the Saguenay's fall to the sea. Downstream some twenty miles, just above tidewater, Lee planned to collect the remaining 210-foot drop behind a second powerhouse to be built where the small Shipshaw River entered the Saguenay. He submitted plans for this second project to the Quebec government in 1925, and the Department of Lands and Forests approved them. When

Shipshaw power project was actually constructed sixteen years later, the layout, including control structures, canal, and power facilities, varied only in the details from Lee's original design.

Lee also contributed his conceptions for water storage to the design of the Saguenay's hydro system. From his first assessment of the river in 1912, the Carolina engineer was convinced of the need to tap the storage potential of Lac St-Jean to regulate the Saguenay's flow. Impounding the lake would roughly double the power-generating capacity of plants downstream and put the project "out into the power markets of the world." Under Lee's direction, Duke's lieutenants secured from the Quebec government the crucial storage rights in the lake. These rights passed to Alcoa in 1926 and were deemed a critical component of the transfer and sale. Alcoa birthed Aluminium Limited two years later in 1928. Thus did Lac St-Jean enter into the control of private enterprise and become integral to the industrial exploitation of the Saguenay River.[2]

In addition to storage in Lac St-Jean, Lee sought additional rights to regulate the flow of the Saguenay's tributaries. As of 1914, he understood that "Lake St. John is not big enough" to store all the Saguenay's waters, and, thus, "how much power you would have [in the hydro system as a whole] would depend solely on what storage could be effected at Lake St. John, or elsewhere."[3] Such a concept of watershed-wide development was certainly ahead of its time. It would be nearly three decades before this vision was realized with Aluminum Company of Canada's impoundment of the Peribonka during the 1940s; yet, on the eve of the First World War, Duke's engineer was already directing the exploration and survey of the rivers Du Chef and Ashuapmushuan, in addition to the Peribonka, with an eye toward the construction of storage dams on these streams.

Given the fragmentary nature of the historical evidence, it is difficult to know with certainty what Lee's specific intentions were in this regard. Where the Peribonka was concerned, at least one scheme called for a storage dam to be built some forty miles below the Peribonka's "forks" (the intersection of the Peribonka and Manouan rivers), with water backed up all the way to this same confluence, creating a very large reservoir.[4] Lee deemed these plans sufficiently advanced, as well as economically necessary, to bring demands for additional storage to the state. In April 1914, Duke's junior partner in Canada, Quebec City timber merchant Benjamin Scott, wrote the Quebec government seeking the right to build "conservation dams" on the Saguenay's tributaries as a means of further regulating the power potential of the Saguenay itself. Because "the flow of the Saguenay varies greatly," Scott argued, "it is absolutely necessary to provide and arrange for the impounding of the waters not only of Lake St. John ... but also the waters of the streams flowing into [the lake] so as to give us a uniform flow throughout the year."[5]

Why Quebec refused to grant the storage rights in question requires some explanation. First of all, the very vagueness of Scott's demands may have doomed the request to rejection by the state. The timber merchant's letter made no mention of particular rivers or lakes and it described no particular impoundment schemes, details that the Hydraulic Service would eventually require to consider such a proposal. In this sense, the request failed to meet procedural standards of the day and constituted an act of audacious presumption. Simply asking for the purchase and/or proprietary control of the several tributaries of Lac St-Jean was akin to asking for the moon.

Moreover, in 1914, since the upper reaches of the Saguenay basin were not yet a part of the province's known hydraulic world, Quebec's bureaucrats could not fathom this application's ramifications. Modern Quebec, of course, is a province known widely not only for the richness of its hydroelectric resources but also for the technological mastery of its provincial utility, Hydro-Québec, which controls them. But the state-mandated corporation was not formed until 1944, and did not come to monopolize Quebec's hydro resources until the 1960s. In the early decades of the twentieth century, the state played but a limited role in the exploitation of hydraulic resources, leaving the capital-intensive task of producing and distributing waterpower-derived energy largely to private enterprise.

Lacking the financial means to build its own power plants, Quebec, nevertheless, assumed important responsibility for their regulation. The Progressive Era brought calls for increased government regulation of private corporations, including electrical utilities, and for increasingly rational use (or 'conservation') of North American natural resources. In this spirit, and beginning in 1910, Quebec's newly formed Hydraulic Service negotiated long-term leases of the Crown's waterways to private firms (in lieu of selling waterpowers outright) and also charged such companies a per-horsepower rate for use of the public domain, creating an important source of provincial revenue. Here was the establishment of an administrative nerve centre to handle waterpower-related data for the benefit of the state; and here was a shift in policy emphasis from short-term financial gain to long-term revenue production.[6]

But the Hydraulic Service was still a fledging bureaucracy – initially, a one-man bureau. And since hydroelectricity was still perceived in Quebec merely as a motive power for pulp and paper production, rather than a resource in its own right, the Service held an office within, and reported to, the larger Ministry of Lands and Forests. The bureau was also limited in its command of information. While surveys of the Saguenay basin's hydroelectric power potential had been conducted over the years in response to inquiries about discrete rapids· or river sections, none had taken account of the exploitation of the full length of the river, let alone the full extent of the watershed. As the Hydraulic Service chief Arthur Amos put it: "The Department possesses ... little in the way of accurate reports relative to this region, and these are but individual, corresponding to one falls or another, without taking a view of the whole."[7] So, for example, following an 1899 offer by Ontarians James Sutherland and Thomas Willson to purchase the lowest two-mile section of the Saguenay, provincial lands surveyor Charles Edouard Gauvin had measured the height and gauged the flow of that part of the river. The previous year, Quebec's superintendent of forest rangers, J.C. Langelier, had reported with much enthusiasm on the Saguenay basin's potential for pulp and paper production, dismissing almost entirely the enormous power potential of the upper Saguenay because it lacked obvious cataracts along its length. Suitably impressed by the lowest falls of the Peribonka (which Langelier compared to the North American birthplace of large-scale hydroelectric production, Niagara Falls), the timber expert, nevertheless, gave no consideration to amassing the Peribonka's powers into a larger development or to exploiting them in the service of the regulated flow of the basin as a whole.[8]

The state's knowledge of its hydraulic resources only got progressively thinner further north. As of 1898, Quebec's territory ostensibly included the entire Saguenay watershed, since the province's northern boundary now roughly followed the 52nd parallel in the vicinity of the Eastmain River; as of 1912, Quebec's boundary shifted northward again to take in the entire Ungava Peninsula on Hudson Strait. However, the industrial frontier moved northward more slowly and incrementally than such facile territorial adjustments on a two-dimensional map. In fact, the great swath of country embracing the upper Saguenay basin remained a hydraulic *terra incognita* to the Quebec state before the era of aerial survey in the interwar period. These subarctic outlands were the familiar domain of Native peoples rather than of the state: Montagnais/Innu hunter-trappers who spent winters in the bush hunting for meat and trapping furs for exchange with the Hudson's Bay Company (HBC), and who journeyed south or southeast by water each summer to the federal Indian reserves on Lac St-Jean at Pointe Bleue or at Bersimis near the mouth of the Bersimis River on Quebec's North Shore. Known aboriginal use of the Peribonka valley goes back at least 2,000 years. By 1912, some 600 Natives were gathering at the Pointe-Bleue reserve each year from the far-flung reaches of the watershed.[9] Another 450 or so were making their way via the Bersimis River, the aux Outardes River, and/ or the Manicouagan River to the reserve and trading post at Bersimis.[10]

None of these 1,000 inhabitants would be consulted by the Quebec state about hydroelectric development in the Saguenay watershed. Moreover, official provincial sources, including the records of bureaucracies charged with forestry and hydraulic exploitation, render the Natives in this vast hinterland all but invisible to the historian. Thankfully, recent oral histories by the Innu themselves permit us, in this and in the following chapter, to begin to restore as visible what was formerly obscured, while a sense of social responsibility compels and obliges us to sketch the outlines of Aboriginal life and livelihood in the Peribonka valley in this pre-industrial period.

Jack Germain (his white/Christian name) was an Innu hunter and his family's migration of 1907–08, the earliest of record, helps to illustrate the nature of Native use of the Peribonka watershed in this era.[11] The Germains' hunting territory straddled the uplands between the Peribonka and Mistassibi valleys and included the Rivière Brodeuse and the Rivière Sapin Croche. Jack himself was born in this vicinity, in the bush, in 1891, making him a boy of fifteen or so as of 1907. In all, eight children and two adults (Jack's parents) made up this winter-season hunting party. Departing from Pointe-Bleue in August, the group traced the western shore of Lac St-Jean in several canoes before portaging the lower falls of the Mistassini and Mistassibi rivers and beginning their ascent of the Mistassibi toward the family's territory. The canoes, likely made of wood and canvas (as the more fragile birchbark craft were being phased out) would have been laden with some 2,000 pounds of store-bought provisions for a subarctic winter, including: 300 traps of beaver size and smaller (made of wood, metal traps being still rare), 2 canvas tents, 15 pounds of flour, 200 pounds of pork, 10 pounds of tobacco, 100 pounds of sugar, 100 pounds of grease (in pails), 25 pounds of tea, and 40 pounds of salt, along with baking powder, soap, some 15 boxes of cartridges for both shotgun and rifle, and candles for light.[12] Knives, needles, fishing line, axes, blankets, and clothes filled out a hunter's list of purchased essentials. Bannock bread would supplement a diet of country food, and the Germains lived off ducks, hare, grouse, and fish (trout, walleye, and pike) during the several-week-long, upstream journey. Working from a main camp and several satellite camps, the family trapped fur-bearing animals from September onward before returning to Pointe-Bleue in June or early July to exchange their pelts for provisions.

Patrick Étienne's memories of the 1914–15 trapping season, when he also was a boy of roughly fifteen, provide additional details of bush life. The Étienne family's territory was situated to the east of the Germain's, in the vicinity of Lac à la Montagne (Lac Duhamel) along the Manouan River. The eleven-member Étienne family – Patrick, his eight brothers and sisters, and their two parents – departed Pointe-Bleue in five canoes, toting two tents and a mass of provisions similar to those noted above. Within a week, the family had ascended the Peribonka to its confluence with the Manouan ('the forks'), from there, pushing up the Manouan for several more days, travelling by portage, pole, and paddle. During the inland voyage, they harvested bear, moose, hare, loon, duck, ouananiche, trout, touradie, and pike. Patrick's mother, Marguerite, also gathered plants for medicinal use during the winter. Once at the main hunting camp, the women remained in place for the autumn season (some three months), while the men departed in two groups – Patrick and his father in one, his brothers in the other – to hunt and trap the territory further upstream. They were gone for several days or a week at a time, tending traplines and hunting, returning to the main site with furs of beaver, otter, marten, mink, fox, lynx, and fisher, and with meat of hare, grouse, moose, and bear, and also various fish (touradie, white fish, walleye, pike, ouananiche).

The family moved further inland to the winter camp in December, where they remained for about three additional months. Again, a main camp, now possibly anchored by a log cabin, served for the entire family, while the males pursued their hunting and trapping forays into the interior, sleeping in the tents and travelling by snowshoe and toboggan. At the main camp, the mother and her girls prepared furs, smoked meat and fish, cut and split wood for heat, and shot or trapped small game. The family returned to Lac à la Montagne for a week or so to hunt and fish prior to descending to the reserve. With the spring breakup, a swiftly flowing Manouan River carried the family by canoe to the forks, a traditional gathering site for Natives returning from the upcountry. Firing their guns into the air to signal their approach, the Étiennes met the Siméon family before completing the downstream and over-lake journey to Pointe-Bleue.[13]

The Siméon family's territory was situated along the Peribonka River well upstream of the forks. The journey to reach it, made in three canoes, could take as long as four weeks; the spring-summer descent to the reserve required a further ten days or so. The family's Aboriginal name, *kwesh-kwan-seguane*, denoted 'one who turned somersaults.' At this date, roughly 1915, Malek Siméon, the headman, was accompanied by his wife, along with their seven girls and two boys, the children ranging in age from infant to young adult. The nearest winter camp was that of his brother, located some five days' journey away.[14] Anne-Marie, Malek's niece, recalls the several-day journey around a windy Lac St-Jean, plying her small paddle in the women's canoe, the men's boats loaded deeply with the winter provisions. Though her mother was French-Canadian, Anne-Marie spoke French only haltingly until, at age eleven, she remained on the reserve for a year to attend the religious school. Otherwise, her youth was spent in the bush. It is bush life that she recalls most vividly in her memoir – in particular, the periods of sharp hunger and the times of plenty with their powerful memories of foods: moose steaks with bannock and hot tea following a successful moose hunt, a pot of ducks following the hunt for wild fowl in the spring, the fatty tail of the beaver, the lean flesh of the hare, the rich meat of the black bear. At fifteen, Anne-Marie married a Native from Bersimis, William Valin, who would hunt on Siméon territory thereafter.[15]

An HBC trader of this subarctic region indicates just how much fish and meat were actually harvested during the annual hunt. Trader J.W. Anderson reported that during the 1912–13 season, the Petawabano family of Mistassini, a party approximating five adults, harvested 430 pounds of moose meat, 6,022 rabbits, 7,300 fish, 100 ptarmigan, 59 beaver, 53 marten, 9 otter, 26 muskrat, 9 mink, 10 ermine, 40 ducks, 3 black bear, and 18 loons. The following year, the family's harvest comprised 3,306 fish, 1,642 rabbits, 67 beaver, 8 otter, 21 marten, 4 mink, 3 ermine, 66 ptarmigan, 140 ducks, 3 owls, 2 hawks, 1 red fox, 2 yellowlegs, 1 gull, 2 caribou, 3 black bear, and 55 muskrat. Aside from the marten, mink, ermine, and fox, everything was consumed as food; and in more dire circumstances, the less savoury furbearers became food as well. Bear and beaver were prized as particularly nourishing and rich in fat. Smaller game such as ptarmigan and rabbit, along with fish, was rightly deemed less sustaining in a winter country, but nevertheless provided adequate sustenance during the inland/upstream autumn voyages to the winter camps, and could be supplemented by salt pork from the trading post.[16]

Other families annually using the watershed of the Peribonka in this era included those by the name of Natipi, Benjamin, Tsernish, Bacon, Washish, Tsheketash, Rock, Fontaine, and Picard, who usually traded their furs at Bersimis (as did those by the name of Canapé, Bellefleur, Rousselot, Paul, Hervieux, Vallée, Vollant, Vachon, Sainte-Onge, and Riverin of the neighbouring Bersimis River); and Régis, Xavier, Raphael, Dominique, and Connolly, who traded at Pointe-Bleue. Departure from the trading posts occurred in late July through August, depending on the length and difficulty of the upcountry voyage. Families returned the following June after a trapping-hunting season of ten or eleven months. Intermarriage between these and other Montagnais families meant that the divisions of territory were flexible over time, it not being uncommon for a son-in-law to secure trapping-hunting lands through marriage instead of inheriting them from his father. By tradition, all Natives were free to hunt for food while travelling through another family's territory. Village school attendance was still rare, as children accompanied their adult parents into the bush. All told, some seventeen to twenty families used the Peribonka and Manouan valleys in the early twentieth century, as their ancestors had for some 5,000 to 7,000 years.

Of course, much change had occurred since the arrival of Europeans.[17] The Montagnais were among the very first of Canada's Natives to make contact with whites. The advent of the French and English fur trades in the seventeenth century enmeshed the subsistence-oriented Innu in a market economy, adding trapping for exchange to hunting for consumption, replacing stone-age technologies with metal ware, drawing the traditional summer gatherings to trading posts, and possibly altering or rigidifying the Native conception of the family hunting territory.[18] In the early twentieth century, birchbark was being replaced by canvas for canoes and shelters, wood by metal in traps, gut by wire in snares, and log houses or tents were increasingly being warmed by metal stoves and chimney pipes. Nineteenth-century colonization of the Saguenay, including land clearing for agriculture and the commercial harvesting of timber, displaced/expelled those Natives whose territories were located in the immediate vicinity of the Saguenay River, of Lac St-Jean, and along the North Shore. The creation of Parc des Laurentides for white sportsmen (1895) similarly rendered a large swath of territory south of Lac St-Jean off limits to Native hunters. Native petitions for land rights moved Ottawa to create federal reserves, including those at Bersimis (1854) and Pointe-Bleue (1856). Canada's Department of Indian Affairs was also successful in establishing a number of Abenaki-Scotch mixed-bloods as farmers in the fertile acreage of Pointe-Bleue from the 1870s forward, which formed the nucleus of sedentary village life.[19] Missionary activity at the trading posts embedded the practice of Christianity in Innu culture by the late nineteenth century.

However, the vast majority of Innu hunting lands remained unaffected and, in fact, never before seen by whites in this era. And by far the greater portion of Aboriginals practised full-time hunting and trapping as their occupation. Some two-thirds of the populations of Pointe-Bleue resided in the bush year-round, returning to the reserve for only a few weeks in summer. Those who engaged in wage work tended toward traditional Native pursuits, serving as hunting and fishing guides during the summer and/or fashioning Native handicrafts (moccasins, snowshoes, etc.) for American tourists who came by train from Boston or New York.[20] At Pointe-Bleue, as of 1910 (if Indian Affairs data can be trusted), hunting/trapping accounted for 74 per cent of all generated income in the community while agriculture contributed only 15 per cent, and pursuits such as guiding and handicraft production provided the remaining 11 per cent.[21] Nearly all of the Natives trading at Bersimis on the North Shore remained full-time hunters, to the consternation of the Indian agent who was determined to see them become farmers and assimilate. "Year after year this band does not make any progress," Agent Adolphe Gagnon bemoaned. "They do not care much for anything else than hunting."[22] Historian Hélène Bédard estimates that, at the outset of the twentieth century, no more than 10 per cent of the population of Bersimis/Betsiamites lived year-round on the reserve, and these were mainly the old and the sick. In short, anyone who was physically able to hunt, did so.[23] Moreover, competing white traders had not yet upset the stable business relationship between the Innu and HBC that dated back to the 1830s.[24]

In short, and despite centuries of interaction with whites, Innu culture remained "subarctic in character, based upon hunting and fishing," wrote American anthropologist Frank Speck of the Montagnais of Lac St-Jean during this period. "Their winter life is passed in the tundra and forests of the interior ... in widely scattered camps consisting of family groups. In the spring they emerge at the coast or at the inland-lake trading posts, living largely by fishing and seal hunting ... to barter their furs for European goods."[25] To the Scottish/Canadian fur trader, as well as the anthropologist, Natives remained the "indispensable primary producers" in Quebec's interior during this "optimum period" of the Canadian fur trade. J.W. Anderson was post manager in 1912–18 for the James Bay Cree of Mistassini, with whom the Montagnais of Lac St-Jean mingled and intermarried. The white man's material civilization had by this point "ease[d] the burden of life" for the Natives of the region, Anderson asserts, "but yet not enough to disrupt their *way* of life."[26]

A boreal Eden this was not. Life was rigorous, at times brutal, in the northern forest. The relatively low carrying-capacity of the interior Shield and the scarcity of animal life dictated the semi-nomadic lifestyle: only by family groups dispersing to distant territories for the winter (what modern urbanites would rightly consider epic journeys) could the Innu hope to earn a living from this environment. And only by exploiting expansive swaths of this terrain (perhaps some 2,500 square kilometres per family), and doing so in sections (harvesting selected stream valleys and leaving others fallow), could a family group hope to sustain an adequate harvest of meat, fish, and fur across many years. Starvation, exposure, and disease remained mortal threats during the colder months. Indian agents of this era describe annual "ravages" of tuberculosis and influenza. Survival required extraordinary resourcefulness and skill in the bush, as well as intimate knowledge of the habits of animals and the layout of the country, particularly the relative position of lakes, rivers, and portages. A woman's life held special dangers. Pregnancy and childbirth in these conditions was, at best, a difficult experience. "It was hard," wrote Anne-Marie Siméon, recalling the birth of her third child, of eight, at the Peribonka's Lac Onistagane in early spring. "I really thought I was going to die."[27] Yet in the early twentieth century, in the Peribonka's uplands, Montagnais life was also relatively autonomous and free: the seasonal migration, from bush to trading post and back again, still maintained a steady annual rhythm. During these voyages, watercourses were of primordial importance as arteries (or "highways" in N.A. Comeau's description of 1909[28]) of transportation and sources of nourishment, the Peribonka being by far the largest of these in the territory used by the Pointe-Bleue Montagnais band. "Home" for these semi-nomadic people of the Saguenay's uplands remained the watery-forested interior rather than a federal Canadian reserve.[29]

What was a homeland to the Montagnais/Innu was a resource hinterland for Quebec. This hinterland was just now being probed and inventoried by advance agents of the state – land surveyors, who, from the turn of the century, made summer or autumn trips into the bush on behalf of Quebec's Ministry of Lands and Forests. They went north to locate and describe the region's natural wealth, especially its timber, which might be harvested to the benefit of the province and its non-Native citizenry. Timber, not hydro, was certainly their primary focus. The late nineteenth century's surge in demand for newsprint in the United States had prompted a surge in the production of Canadian softwood and wood pulp. In the vast and forested Saguenay region, Quebec City–based English-speaking entrepreneurs like Benjamin Scott and William Price had scrambled to acquire cutting rights (or timber "limits") in the vicinity of the basin's river arteries, including the lower Peribonka.[30]

The state abetted timber cutting by scouting ever northward for forest resources. By canoe and by foot, Quebec surveyors ascended, or "scaled," the rivers of the Canadian Shield, returning to Quebec City to file their reports. Surveyors William Tremblay and Georges Leclerc, for example, ascended several branches of the Peribonka in 1911 where they found riverbanks "fairly well wooded with white birch, black spruce and balsam fir"; the following year surveyor Paul Joncas described "fine pulpwood" (mainly black spruce) in the valley of the Manouan River, as well as larger "grey and white spruce from eight to fifteen inches in diameter, white birch from five to seven inches." Joncas also noted the region's game animals and (in a rare occurrence in these surveyors' reports) the hunter-trappers who relied on them. "During the winter the Indians camp [in this vicinity]," he wrote, of a patch of lakes in the Peribonka's highlands, noting also the Natives' practice of hanging animal remains in the trees to propitiate the animals' spirits. "Skulls of beaver and bear with bones of lynxes and otter can be seen at all their camping places." It is clear that hydroelectric power was not the focus of such state inventories. Surveyors sought "merchantable timber" or wood that would be "suitable for making pulp." The exploration of a hydrographic basin was carried out "with reference to the pulp and paper industry": individual waterfalls were described when their exploitation might serve a future saw mill or pulp mill, and rivers were noted as to whether they were "floatable for logs." But watercourses, a number of them not yet properly mapped, were also not yet assessed in the context of the flow regime of the entire watershed or with consideration of water storage to serve power plants to be built well downstream.[31]

So, why was it that Quebec's civil servants refused to allow businessmen to dam and harness the Peribonka River just prior to the First World War?

The Native presence in the upper Saguenay watershed most certainly had nothing to do with it. Rather, in the case of the Peribonka, the Americans' conceptions of northern resource exploitation exceeded the provincial government's rational understanding of its own domain. The Hydraulic Service was unwilling to part with what it did not yet fully know or understand and so proceeded with caution rather than haste. "In principle," the minister of lands and forests informed Duke's agent, "the government is in favour of all hydraulic works tending to develop the potential of running waters." But given the paucity of information available on the tributaries of the Saguenay, including that supplied by the investors themselves, it would be impossible to make concession of such waters or waterpowers at this time. "The government would be disposed to look favourably upon projects considered one by one ... but ... naturally, the nature and layout of these works would have to receive the approval of the government ... and under condition that the government could take possession of these dams at any time, while reimbursing you for their cost."[32] Evidently these were excessively pro-hibitive terms for James B. Duke. There was no further correspondence with the Quebec authorities on the development and utilization of the Peribonka River until the eve of the Second World War.

Industry's demands did, however, rapidly move the state to further its own knowledge of the Saguenay watershed, and a study of Lac St-Jean marked the starting point in this regard. Early in 1915, Duke's lieuten-ants laid out for the Hydraulic Service in explicit detail their plan to use and enlarge the lake as an industrial millpond. Hydraulic Service chief Arthur Amos, now fully cognizant of the intentions of the Americans, called for a technical study of the lake in order to make sense of these large-scale demands and to enable the state to effectively negotiate with the Americans. How much power would the Americans gain from the benefit of storage? How many feet rise in lake elevation could the devel-opers be permitted without excessive damage to agricultural interests? How much money could the province legitimately tax the Americans for use of the public domain?

We should explain that whereas the Hydraulic Service formed Quebec's original administrative centre for the management of the prov-ince's waterpowers, it was the Quebec Streams Commission that acted as the province's technical advisory in the field of hydraulic engineering. Also formed in 1910, the Streams Commission was originally created to "devise just and practical rules respecting ... the preservation and man-agement of running waters." Granted additional powers as demanded by each project, the three-member, Montreal-based Commission was eventually responsible for the construction of expansive, government-operated storage reservoirs, the first of which (the so-called 'La Loutre' and later 'Gouin' reservoir) was built high in the valley of the St-Maurice River beginning in 1915. Such reservoirs yielded substantial revenues for the province while serving the power interests of private industry.[33] For our purposes here, we note that the Commission also collaborated with the province's Hydraulic Service to assess the demands of potential investors. In April 1915, Amos requested the civil engineers of the Streams Commission to make a careful survey of the Saguenay basin to aid in his negotiations with the Duke interests, a study that should begin with the "possibility of storing water in Lake St. John."[34]

The study touched on the lake itself, as well as its tributaries and surrounding farmland. Concerning the lake, the Commission's November 1915 report confirmed what the Duke interests already knew, that Lac St-Jean made an "ideal reservoir for the control of the waters of the Saguenay." The report described what lands would be submerged at what lake levels and with what benefits for power production. Upstream, the Commission began to measure the flow volume of the tributaries in order to assess the role played by these rivers in the control of the Saguenay itself; such gauge work, now begun, would continue thereafter. The Commission's engineers also determined that the tributaries would be affected by a rise in lake level, since several rapids and waterfalls on the lowest reaches of the Ashuapmushuan, Mistassini, and Peribonka rivers would be inundated by the lake's impoundment. Finally, the state came to confirm for itself that Lac St-Jean could only hold so much water; to further raise the regulated flow of the Saguenay and, thus, the generating capacity of its power plants would require "additional reservoirs in the basins of the tributaries."[35]

Clearly, additional study of the Peribonka was called for. The 1915 study became the first of several by the Commission that expanded the state's knowledge of the basin's hydraulic potential in subsequent years.[36] During the summer of 1917, Streams Commission engineer T. Toupin made an initial investigation into storing water in the Peribonka basin. Toupin returned from his two-month sojourn with notes on the valley's topography and on its larger lakes. In the summer of 1920, the Commission pursued a more exacting study of the river's profile under engineer Éloi Duval, carefully mapping the streams first thirty-one miles above the village of Honfleur (at the Peribonka's last rapid where the river meets the level of Lac St-Jean), establishing benchmarks along the route, drawing plans of each major rapid and assessing each one for its potential to generate electrical power. By 1922, the Commission had assembled its acquired knowledge of all of the Saguenay's tributaries in a single report. In this same publication, the Commission set out its own plan for the storage of water along the Peribonka River. Reflecting the state of the hydraulic art, the plan called for just one or perhaps two dams to be constructed at strategic points along the lowest thirty or so miles of rapids and falls (in fact, precisely where Alcan would do so in the 1950s: at Chute-à-la-Savane and Chute-du-Diable), in order to amass some 220 feet of the river's drop or head and produce some 300,000 horsepower. The scheme promised to fully triple the minimum year-round flow of the Peribonka, from some 4,000 cubic feet per second to some 12,000 cubic feet per second.[37]

As it happened, neither the state nor the Duke interests developed the Peribonka basin in this era. However, Duke's company did gain provincial concession of storage rights in Lac St-Jean in 1922. Duke-Price Power Company then built a series of dikes and dams and an enormous power plant on the upper Saguenay River at the place called Isle Maligne. Completed in the summer of 1926, this plant backed water directly into a broadened and now-regulated Lac St-Jean to benefit from the lake's enormous storage capability.[38]

Still, the state didn't rest in its acquisition of knowledge of the boreal hinterland. During the 1920s, the Surveys Branch of the Department of Lands and Forests was engaged in rolling back Quebec's resource frontier by extending south-north 'meridians' (lines of longitude) into the northern sector of the province and then laying down east-west 'base lines' (parallels of latitude). Extending the grid permitted surveyors to explore and make accurate maps of the northern waterways, binding them to the province's known world and opening them for future development. In a veritable flurry of activity from 1925 to 1932 (1925 being the first year Quebec surveyors began using aerial photography to supplement ground operations), surveyors mapped an impressive 51,000 linear miles of watercourses, including the upper branches of the St-Maurice and Ottawa rivers, several basins of Quebec's North Shore (the aux Outardes and Manicouagan), and even several rivers of the Abitibi region (the Rupert, Broadback, and Nottaway) that drain northwestward into James Bay.[39]

In 1926, the Peribonka joined this distinguished list when Lands and Forests ordered a complete traverse (or survey) of the river to its northern-most source. Surveyor Louis Giroux headed the first and formidable stage of this effort, which involved a total of three trained surveyors, 126 men, some 50,000 pounds of provisions, and an estimated three summer seasons of work to map the basin as far north as the not-yet-fully-surveyed 51st parallel (just north of the reservoir Lac Manouan). "The Peribonca presents a wide field for the development of hydro-electric energy," Giroux concluded, "its immense basin permits the control of waters to insure uniform flow." Final phases of the traverse, led by seasoned surveyor H. Bélanger, took advantage of the extraordinary access to the bush offered by the airplane. With only one companion, Bélanger overflew the previously surveyed region and then charted the upper reaches of the Manouan River. Although it is difficult to determine with precision the actual dates and specific geographic coverage of these expeditions, we can nevertheless summarize their achievement by quoting the director of surveys in a report penned some years after the fact: "The Peribonka was surveyed from 1926 to 1929," wrote Georges Côté. "Starting from a point on this river near parallel 49, up to its source, north of parallel 52, its course is practically North-South." Within several years (by 1934), provincial cartographers had quite accurately drawn the Peribonka and its several feeder streams onto the provincial map.[40]

The timing was propitious. In fact, Quebec's waterpower experts were quite eager to put such information to good use by the late 1920s. In June 1926, Duke-Price Power Company (builder of the Isle Maligne power station) closed the gates of its spillways to raise the level of Lac St-Jean – without first compensating the area's farmers for thousands of acres of soon-to-be-flooded land. Through 1927, the resulting outcry and protest by the region's rural populace gained anger and strength, and then spilled into the province at large. A Montreal journalist expressed his shock at *"la tragédie du lac St-Jean"* and the name stuck. The events of 1928 only exacerbated the earlier controversy: heavy spring rains fell high in the Saguenay watershed, augmenting the annual spring freshet and flooding the Saguenay valley below. Lac St-Jean rose higher than at any time in recorded history, inundating whole villages and washing out bridges. For local farmers, already irate with American corporate interests, *"l'inondation de 1928"* [the flood of 1928] was but the extension of the earlier "tragedy" and further proof of the callous abuses of foreign industry in its exploitation of the province's natural resources.[41]

The "tragedy" of Lac St-Jean would move the state's waterpower bureaucracy to revisit and advance their plans to develop the Peribonka

River. As early as the autumn of 1926, within just several months of Lac St-Jean's impoundment, a young engineer with the Hydraulic Service suggested a solution to flooded farmland around the lakeshore. Junior civil engineer Raymond Latreille suggested to his superior Arthur Amos that creating reservoirs along the Peribonka might permit the Americans to lower the lake level, while still maintaining high constant water flow through their turbines and thus a high level of electrical generation. In the wake of the 1928 flood, Latreille returned to this notion with new urgency. Quebec's hydraulic engineers were more reticent than regional farmers to assign blame for the flood to the industrialist's manipulation of the environment. In fact, after careful study, representatives of both the Hydraulic Service and the Quebec Streams Commission concluded that Duke interests' partial excavation of the lake's outflow channel during dam construction, which was done in the service of power production, actually meant that the lake's floodwaters were purged more quickly than they would have been prior to dam building. *Force majeure*, rather than Duke-Price Company, was largely responsible for the 1928 flood. Still, Quebec's engineers were sympathetic to the farmers' woes and also determined to ameliorate the situation within the bounds of their skills and authority. In October 1928, Latreille urged his chief to consider a Peribonka scheme. Why not pursue reservoir construction in this remote basin as a means of flood control for the Saguenay's more populous regions, Latreille argued, and such a proposal might well be attractive to Duke-Price since storage meant more power for the reduction of aluminum and the making of paper from wood pulp.[42]

Amos evidently conveyed these suggestions to the Quebec Streams Commission, of which he was also a member.[43] As early as 1927, in the midst of the furor over flooded farmlands around Lac St-Jean, the Commission began an investigation into the storage of water in the Peribonka basin, aiming also at the hydroelectric development of the lower river; by 1928, the year of the flood, these studies were in full swing during both summer and winter seasons. It is not clear whether the Commission saw the project as one that would be funded by government, or one to be carried out by Duke-Price Company. In any case, by 1929 the Streams Commission identified the head of Grand Rapide (fifty-three miles above the Peribonka's mouth) as "the most favourable location for the construction of a dam," a dam to be built partly of concrete and partly of earth and to cost an estimated $5.5 million. Its reservoir would extend upstream for nearly sixty miles, back beyond the forks of the Rivière Manouan, and would double the Peribonka's year-round flow to at least 8,000 cubic feet per second. Steep banks in the reservoir's vicinity would keep the lateral flooding of land to a minimum; and, in any case, since "the entire territory is neither surveyed [for agriculture] or improved," the flooded area was deemed "unimportant" to provincial interests.

Provincial interests, in short, embraced farmers and factory workers and, in general, those who were sedentary and white. Provincial interests did not include mobile hunter-trappers of Aboriginal heritage who held no firm, documented title to the land.

A second project was also proposed: the creation of a reservoir high in the valley of the Rivière Manouan (very close to where Alcan would build its storage dam in the 1940s), territory that was just now charted by the surveyors and cartographers of the Department of Lands and Forests. Such was the site's remoteness that the Commission used planes to access it. The planned concrete dam and accompanying earthen dike would be built at the outflow of Lac Opitoonis and cost some $1.1 million. The reservoir would fill and slightly enlarge the natural chain of lakes upstream in the vicinity of so-called Lac Péribonca (also called Lac Manouan). In tandem, the two projects would raise the regulated flow of the river to around 12,000 cubic feet per second to serve power plants well downstream, while also generating some 400,000 horsepower in the lower Peribonka itself (or roughly the amount of power produced by the huge Isle Maligne station). Finally, the scheme served the interests of flood control. As the Commission understood it, "storing water in the Peribonka River would diminish the maximum flow into Lac St-Jean, thus mitigating in a measure the high waters of the lake."[44]

Why neither the state nor private capital pursued these plans in the early 1930s is not specified in the historical record. Certainly, the Depression played an important role. By 1931, the market for aluminum had collapsed, and electrical power from the existing Isle Maligne plant, and also the newly completed Chute-à-Caron plant downstream, ran largely to waste. Thus the Depression removed any incentive, for industry at any rate, to augment the current power supply with the establishment of additional water storage elsewhere in the Saguenay watershed.[45] "We had surplus power," Aluminum Company of Canada (Aluminum) president R.E. Powell stated bluntly of the 1930s. And in this period of financial "suffering ... when no one would buy [our] products ... stocks of the metal became so burdensome that Alcan resorted to barter for coal, oil and almost anything it could turn into money." Only the revival of business beginning around 1937, due to rearmament, would stimulate the expansion of Aluminum's production capacity of aluminum as well as the power facilities required for the same.[46]

So the Peribonka River still flowed freely in the late 1930s. Although industrialists had eyed the energy resources of the valley for several decades, and the Quebec state had responded to private capital's demands by making efforts to understand their scale and their nature, no dam or reservoir checked the river's course. No construction project was imminent anywhere along the river's 300-mile length.

What had been accomplished then? The Quebec state was now familiar with the course and hydraulic contribution of the Saguenay's tributaries. By the time of the Great Depression, Quebec's hydraulic bureaucracy had itself fashioned specific plans to develop the Peribonka for its energy potential as well as to check the floodwaters in the basin as a whole. And whether or not the state acted on these plans, the acquired knowledge would prove useful. If and when industry renewed its interest in the basin, the state would be ready and fully able to negotiate. In this sense, the events that unfolded as early as the First World War would shape the developmental outcome of the Saguenay watershed during the Second World War to come.

The Peribonka's prehistory also provides a valuable illustration of the Province of Quebec hydro policy in the early decades of the twentieth century.

Progressive reforms were changing that policy in significant ways, although Quebec was somewhat slow to regulate its hydro industry. From the North American inception of hydroelectricity in the 1890s to the year 1910, waterpowers in Quebec were sold outright, in perpetuity and on the cheap, to interested capitalists, even while neighbouring Ontario nationalized hydroelectricity (1906) and the Roosevelt administration in the same year greatly restricted waterpower development in the United States. The establishment of the Hydraulic Service and the Quebec Streams Commission in 1910 inaugurated an era of real reform and regulation in the hydro sector in Quebec. The Service set about thoughtfully considering private industry's demands and making efforts to accrue for the state long-term financial remuneration by leasing, rather than selling off, the Crown's rights. The Commission collaborated with the Service in the task of technical studies, all the while engaged in the increasingly efficient use of waterpower by means of state-funded construction of storage reservoirs beginning on the St-Maurice River.[47]

The staff of these organizations also grew increasingly sophisticated and foresighted as practitioners of the hydraulic arts. Prior to the First

World War, the state's efforts to understand its own hydraulic North amounted to disjointed probes, lacking understanding of what it meant to amass the power of multiple rapids and waterfalls by exploiting the river's full power reach. By the early 1920s, spurred on by private-sector visions of profitable development, rational planning replaced mere reconnaissance and the state was employing skilled technicians to assemble accurate maps and flow data for future use. The Quebec state's motive behind assessment of the hydraulic frontier also shifted, from examination of hydro-timber potential, often in reaction to specific requests of industry, to systematic data-collection and planning with regard to hydroelectricity per se. In this way, the Progressive Era's penchant for scientific management reached Quebec's North. In this way, too, Quebec's civil servants, charged with the management of the province's waterpowers, brought government up to speed with private industry in its knowledge and its development conceptions of a remote northern watershed.

It is evident that Quebec's hydro reforms had their limits, as well as their strengths. The most important limit was likely financial. Early twentieth-century Quebec was a capital-needy province. Sharp-eyed historians have noted that the cost of Duke's Isle Maligne dam exceeded the total annual revenue taken in by the Quebec government in the mid-1920s when that dam was built.[48] Limited state resources meant relatively small staffs, possessing limited knowledge of their own territorial domain. It was the American engineer William States Lee who first crafted plans to develop the hydroelectric potential of the Peribonka River in 1914, prompting the state to begin a study of the river's flow and profile. Again, in the late 1920s, industry prompted state action, when the damming of Lac St-Jean provoked a regional political crisis and moved the province to look at the Peribonka as a means of flood control. In both instances, the state's efforts were reactive, instead of proactive. In both, Quebec's hydro experts were forced to play catch-up with foreign investors who still outspent and temporarily outpaced state functionaries.

2

Lac Manouan

The Great Depression quashed any additional hydroelectrical development of the Saguenay watershed through the mid-1930s. By this decade,
Lac St-Jean controlled the flow of the Saguenay River; and in tapping
this regulated flow of water, Aluminum was producing electricity at Isle
Maligne (1926) and at a partially completed power plant at Chute-à-
Caron thirty miles downstream (1931). By the 1930s as well, Quebec's
Surveys Branch of the Department of Lands and Forests had mapped
this territory with the aid of aerial survey. Quebec's Streams Commission,
prompted originally by private enterprise's contemplation of reservoir
construction in the Peribonka valley and also by controversial flooding in the Lac St-Jean area (the tragedy of Lac St-Jean), had suggested
schemes for the basin's regulation and control. In short, what had
been the uncharted domain of Native trappers had, by the 1930s, been
brought within the purview of the Quebec state.[1] Rearmament would
trigger the next phase of the basin's development.

The Natives themselves, meanwhile, were still quietly and thoroughly
using the land in question throughout the interwar period. Looking
closely at several annual migrations of this era helps us to understand
what, if anything, had changed in Aboriginal use of the Saguenay's
uplands since the turn of the century, as well as what had remained
essentially the same.

Baso (Joseph) Boivin (an older brother of Louis-Georges Boivin,
whose migration of 1938–39 is summarized in the Introduction) was
twelve years old during the 1923–24 migration into the interior. His
father was Charles Boivin, an Attikamek hunter based at the Obedjiwan
reserve at the headwaters of the St-Maurice River. Here, logging was rendering hunting and trapping increasingly difficult. This may explain why
Charles, back when he was seventeen, had journeyed with his parents
and family by canoe to Lac St-Jean (where Charles would be accepted as

a member of the Montagnais)[2] and why Charles Boivin and family now trapped the far more remote headwaters of the Manouan, Bersimis, and aux Outardes river valleys.

In the first days of August 1923, father Charles, brothers René, Charlot, and Baso along with the families of Michel Dominique, Barthélemy and François Germain, Paul St-Onge, and perhaps others hired a buggy from Pointe-Bleue to nearby Roberval. There they took on provisions from independent trader Léon Roy before boarding the lake steamer *André Daniel* (supposedly at no charge to the Natives) to the farming hamlet of Honfleur. The group of twenty-five (including nine young children) proceeded by three hired cars to Chutes McLeod along the Peribonka, twelve miles above Honfleur. Here the Boivins off-loaded provisions into four canoes: 1,000 pounds in each, which included 400 pounds of flour per person, 15 pails of grease, and supplies of tea, coffee, sugar, raisins, prunes, rice, and beans. Before beginning the lengthy upstream journey, the flour had been transferred into fifty-pound sacks stored, in turn, in wooden boxes.

Traps, snowshoes, and other tools had been cached during the spring before. Nevertheless, transporting the mass of provisions was time-consuming, heavy labour: rapids that could be ascended by pol-ing required three trips for each canoe; those that called for a portage demanded eight or even nine trips across the trail. Particularly rigorous was the month-long journey to skirt Passe Dangereuse ('the long rapids'; to Natives: 'les rapides Kapitatshutsh') by ascending the Rivière Serpent and Rivière Brodeuse. Just the first two of the many portages, at three miles each, required a full week's work. The strenuous work was too much for one of Michel Dominique's young sons who "pushed too hard [and] broke a vein" and was buried in a Native cemetery in this vicin-ity.[3] Crosses marked the bodies of a half-dozen other children who had perished from illness or injury in years passed, giving the name *la passe des morts* [passageway of the dead] to this section of the route. The work remained rigorous even when the party rejoined the Peribonka, as river currents and/or headwinds made progress difficult. "It was hard for the arms," Baso recalled.[4] While ascending, passing through the territories of Xavier, Siméon, and Natipi, the Boivin party snared hare at each camp-site, shot grouse and duck, and netted or trawled for fish. Baso's father smoked some of the fish to preserve it. They killed a young moose in the vicinity of the future Lac Peribonka/Passe Dangereuse reservoir. The other families gradually went their separate ways: François Germain at the Rivière Serpent, Michel Dominique at Lac Onistagane, Barthélemy

Germain at the Rivière à la Carpe. The four Boivin males continued north, up the Savane River to the confluence of Rivière à Michel.

The journey to this point had taken three months. It was now late October: ice was forming in the river's eddies, snow fell regularly, and the fur-bearing animals were ready for harvest. Upon arrival at Rivière à Michel, the Boivins set up a main winter camp with a large square canvas tent and metal stove. They prepared snowshoes and toboggans from the surrounding birch woods, and then began to trap and hunt in earnest. Baso was paired with his father, while the two older brothers were by themselves. For several weeks at a time, these partners tended traps along lengthy waterway circuits, tracing lakes and tributary streams deep into the interior, sleeping each night in a small tent, eventually returning to the main camp to unpack furs and restock provisions. They harvested muskrat, marten, mink, fox, otter, and beaver; netted fish under the ice; snared hare; shot moose in the forest; and further north, as the trees thinned, hunted woodland caribou. The hunters disposed of animal bones with the proper respect, hanging skulls in the spruce trees with a bit of string or wrapping the bones in cotton cloth before placing them in the trees. "If you don't do that, you are not 'lucky'," Baso recalled. By springtime, the Boivins' travels had taken them well north and overland into the headwaters of the Bersimis, Manicouagan, and aux Outardes rivers, where they met Native families from the Bersimis post. By April, back in the valley of the Savane, they were caching their traps, toboggans, snowshoes, and smaller hunting canoes on an island (to avoid possible destruction by fire during the summer) and packaging their furs in eight tight bundles for the return journey downstream.

The Boivins descended swiftly in two canoes, accomplishing in ten days' time what it had taken three months to achieve during the autumn upstream journey. As the rivers were swollen by meltwater, they took care to portage the major rapids, lest their canoes upset and the occupants drown. As they were in a hurry, they did not bother to fish or snare small game. Instead they shot geese, ducks, and muskrat that crossed their path, and otherwise sustained themselves on dried and crushed moose and beaver meat, prepared in previous months by Baso's father, mixed with fat. The party saw fresh signs of moose, but killing such large game was out of the question, as the meat would go to waste. At the Manouan-Peribonka forks, the Boivin party met several fur buyers (including Aldège Tremblay and 'Tit-Rouge' Drolet of St-Félicien, as well as Clément Dufour) who had ascended the river by motorboat in hopes of purchasing Native furs. It was the traders' dubious practice, Baso remembered, to sell the Natives liquor in order to get the furs at bargain prices. Baso's father avoided the trickery of these "robbers" by passing right on through. Other headmen, showing less willpower, might spend a week at the forks where "they drank practically half of their furs."[5] Some distance further down, the Boivins used the telephone at the fire warden's station to call for a cart or automobile: French-Canadian farmers Joseph or William Desbiens were happy to oblige. At Honfleur, the Boivins awaited the steamboat with several other Aboriginal families – Dominique, Germain, Sylvestre – while enjoying the provisions of the local store. Here the atmosphere was festive. When the steamboat arrived, Captain André Daniel took the Natives across Lac St-Jean to Roberval, passing close enough to Pointe-Bleue to sound the ship's horn in greeting; the Natives themselves stood on the ship's bridge waving red kerchiefs to their relations on shore. At Roberval, the Aboriginals debarked into motorcar taxis to return to the reserve. In all, the Boivins' journey spanned eleven months and probably over 800 miles via canoe or snowshoe.[6]

David Philippe, of Abenaki-Canadien descent, recalled being part of the hunting migration of 1926–27 as an unmarried man of twenty. Like other mixed bloods, the Philippe family lived by farming on the reserve as well as by participating in the tourist trade. David's father, a canoe builder, guided for the prestigious Triton Fish and Game Club located at the height of land between Quebec City and Lac St-Jean. David himself learned to hunt with his older brothers and his brother-in-law Emery Connolly, a full-blooded Aboriginal. At the age of sixteen or so, he began hunting full-time in the most remote quarter available to the Montagnais, the Peribonka valley; 1926 would have been David's fourth or fifth year as year-round hunter. Interestingly, it was also the year of completion of the Isle Maligne generating station, the largest installation of hydroelectric power on the planet. Not twenty miles away, at Pointe-Bleue, David and his brother Bazo and three other young men of the Boivin family, toting two tents and accompanied by a dog, set out in the month of August: by cart to Roberval, by the steamer *André Daniel* from Roberval across Lac St-Jean to the village of Honfleur at the lowest rapid of the Peribonka, and then via buggy again up the river to Chutes McLeod. From this point, the men proceeded by canoe, living off ducks, geese, and muskrat, and also setting their nets each evening for whitefish, pike, and carp.

Within six days of canoe travel, paddling and portaging (and poling – three trips through each rapid), they reached the forks of the

Manouan. Within another hard week of travel that included miles of portaging up the Rivière Serpent and Rivière Brodeuse in order to skirt the long rapids of Passe Dangereuse, they returned to the Peribonka at what is now Lac Peribonka/the Passe Dangereuse reservoir. While the Boivin group required a month to skirt Passe Dangereuse, it is likely that the Philippe group managed it in a week due to the absence of women and children and associated baggage: each portage required but six trips rather than nine, and the all-male group had fewer mouths to feed. Nevertheless, this was grueling work: "It was hard," David recalled. "A heavily laden canoe, sixteen feet long and almost one thousand pounds inside, it's hard to manoeuvre in the water. We had no motor back then. On the portages, we made six or seven trips. That's fourteen trips across."[7] Poling and paddling onward for another week, they reached Lac Onistagane, where the wind held up their journey for three days; the Bacon and Dominique families were also camped here awaiting a change in weather. David and his companions cached some fifty pounds of their winter flour on a scaffold. Another five days' travel beyond brought them to the mouth of the Rivière Savane; and several days further upstream, in the vicinity of Rivière à Michel, David left his companions and proceeded alone up a small stream to Lac Benoit. In this area, and by invitation of the Boivin family, he intended to trap and hunt during the winter months. By this point, the men had encountered some twenty-eight other Native canoes in the Peribonka basin. Before summer, David would visit with several other groups of hunters, including his travelling partners, the Boivins, and the family of Berthélemy Germain.

The young man fashioned a pair of snowshoes, which he cached along with lard, flour, and other necessities at a main tent camp. He then trapped and hunted beaver by canoe until freeze-up. As there were other trappers in the vicinity, David took care to mark those beaver houses that he found so that others would leave the animals to him. Thereafter, via snowshoe, he worked traplines along the tributaries of the Savane. We should emphasize that while David and his companions used lake steamer or horse-drawn buggy at the outset of their trip, travelling conditions in the bush remained primitive, and navigation was conducted solely from memory. "We had nothing in those days," David recalled of the 1920s, "no motor, no map, not even a compass. It was the Indians who told us how it was done. If we came upon new territory, as at the headwaters of the Peribonka, we walked and we found the lakes and we gave them names. We knew whether this one or that flowed into the

Bersimis, or to Lac St-Jean or to the Manicouagan because we hunted right up to the height of land."[8]

David recalled a difficult early season. While the Boivin brothers were catching beaver, mink, otter, and marten ("beaucoup de belles martres") in good quantity, David himself had little luck early on. "I saw that this [area] had been gone over, it had been hunted the previous year. The Boivins hadn't sent me to a good territory."[9] In addition, he had picked up a case of body lice at one of the campsites during the upstream journey and was hard pressed to rid himself of the torment, but tried, by boiling his clothing in a small cooking pot. Trapping solo also added to the season's rigour: he often found himself preparing beaver skins until the wee hours of the morning. David roamed further afield onto the lands of the Siméon and Dominique families, where his luck began to change. With the help of his dog, he tracked and killed a large female moose bearing plenty of fat, one of two moose he would shoot that winter. Trapping a birch-covered mountain, he also harvested trappers' 'gold' in the form of a dozen lynx, each worth $80; and he caught additional beaver as well as several marten and otter. With the arrival of spring, David had ranged southward and was hunting at Lac Trippe d'Ours (Bear Blood Pudding Lake) and Lac à Michel, where he shot five caribou in the company of Berthélemy Germain. The two hunters remained long enough to smoke the meat before starting south on the downstream journey.

David and several members of the Germain and Sylvestre families – a total of twelve adults and two children – descended the Savane and Peribonka by toboggan, towing their canoes behind. Michel Germain became sick to his stomach and left a toboggan behind that was loaded with meat. To make up the shortfall in provisions, the others shot ducks, geese, and muskrat, snared hare and made bannock from the cache of flour left along the Peribonka that autumn. The ice had gone out at Lac Onistagane, so they proceeded from there by canoe. David paired off with François Savard to shoot wildfowl. Savard winged a goose, using the tethered live animal as a decoy to attract others. By the time the two young men met up with the others, they had plenty of meat to share. Skirting Passe Dangereuse via the Brodeuse and Serpent rivers, the group met the families of Jean-Baptiste Dominique and Malek Siméon. Eventually reaching the forks, the men fired guns into the air in salute. At this traditional gathering spot, David lingered for two days and sold his beaver pelts to Mr Moar of HBC. Two days' travel south of the Forks brought them to Chutes McLeod, where a small sawmill marked the beginning of French-Canadian civilization. David hired a truck ("un petit Ford") to take him the rest of the journey by road via the village of Alma. Following a difficult start, this trapping season yielded considerable profit: David grossed $3,000 for the furs from this run.

Joseph Natipi recalls the migration of 1927–28, when he was twenty-one years old. The Natipi territory lay along the upper Peribonka River, adjoining the hunting lands of the Mistassini band in the valley of the Témiscamie and Eastmain rivers. Joseph was born along the Big Otter River in this vicinity. In August 1927, his father, Paul, led a group of five adults, including Joseph's mother, Mathilda, and three small children. Departing early in the summer season because of the long voyage upriver, they made their way by truck around the shores of Lac St-Jean to the rapid Cran Serré, and from there began the lengthy canoe trip up the Peribonka River. Within two weeks, they reached the forks of the Manouan; another three weeks of hard portaging brought them around Passe Dangereuse via the Riviére Serpent and Brodeuse, and back to the Peribonka. Low water slowed the journey, as did the family's mass of provisions, which included 2,000 pounds of flour and nine pails of grease. With no time to fish during these long days of labour, and bannock being an insufficient source of energy, the Natipis lived off the *petite chasse* of hare and grouse. Once on the Peribonka again, above the Passe, they could place their fishnets each evening to catch pike, whitefish, and carp. Although they saw signs of moose, they did not hunt such large game at this stage of the voyage; it would require too much time to smoke the meat, which otherwise would go to waste. They did shoot a bear somewhat further upstream. Four more weeks of upstream travel brought the group to Lac Natipi, several days' travel above the mouth of the Savane River, where they built a log cabin as winter camp. All told, the upstream journey took over two months.

Over the following month, the Natipis trapped the lands in the vicinity of Lac Natipi, catching mink, marten, beaver, otter, and muskrat. They also constructed snowshoes and toboggans in preparation for winter. Joseph and his brother (also named Paul) travelled north by snowshoe for several days to trap the lands in the vicinity of Lac Nipissi and the Eastmain River, where they remained for about six weeks. The brothers returned to the cabin at Lac Natipi in time to celebrate the Christmas/New Year's *fêtes* with their parents and younger siblings. As provisions were low, the family took advantage of the two-week visit by the older brothers to bulk up their supply of fish (touradie) by fishing beneath the ice. As winter turned to spring, the men (father and older brothers) made several hunting forays by canoe along the Sugar River (Rivière au Sucre). They caught muskrat, otter, and beaver, but killed no larger game. Caching the traps at Lac Natipi, along with other belongings not required for summer, the entire family then began the descent of the Peribonka. They found no fur buyers at the forks, as most of the Aboriginal families had passed by some time before. At Lac Tchitogama they found two buyers and chose to sell their furs to trader Joseph Casey. From Tchitogama they portaged overland to the Rivière Bourget and downstream to the small farm village of St-Ambroise. Here the Natipis hired a horse and cart to take them by road to the town of Chicoutimi where they would spend the summer.

Jean-Baptiste Dominique recalls journeying with his family into the upcountry in the 1929–30 season at the age of twelve. His father, Michel, trapped a territory along the upper reaches of the Peribonka River. The group of nine, including two young children, made their way up and beyond the forks, using the Rivière Serpent and Rivière Brodeuse to skirt the violent rapids of Passe Dangereuse. Along the way, they snared hare, shot partridge, and fished for pike, walleye, and trout to sustain them. It required five weeks of upstream work to this point. Once at the site of what is now Lac Péribonka (Passe Dangereuse) – formerly a winding and grassy-banked river, and an ideal habitat for muskrat – they trapped these wetland rodents as they continued upstream to Lac Onistagane and the Rivière Savane. It was well into October, an eight weeks' journey north of Lac St-Jean, before the family made their winter camp in the valley of the Savane. Here they remained until March.

Throughout the winter, the men set out in small groups to tend traplines from several satellite tent camps, harvesting mink, marten, otter, fox, and pine marten; the women and children remained in the main camp, also a canvas tent, to prepare the furs, cut wood and gather food. Woodland caribou meat helped to sustain the family through this period. In mid-March, the Dominiques began a gradual descent via the White Mountain River, making their way as far south as Lac Manouan, where young Jean-Baptiste became ill. "I do not want to hold you up," the boy told his father. "If I am going to live I will live and if I am going to die then I will die."[10] But the family remained at Lac Manouan for a month so the boy could recover. An old woman daily fed him an evergreen infusion (made from *ka kauatsh*, which resembles the fir and grows among balsam), and eventually hare intestines, and a tea of wild cherry (*apueiminan*) and beaver kidney. The family spent several more days at the HBC outpost at the outlet of Lac Manouan. With the boy's health recovered, the family continued its spring descent along the Manouan River to the forks. Here they traded some of their pelts with white traders before completing the journey to Pointe-Bleue by motorized canoe, a technology that reduced the week-long journey to one-and-a-half days.[11]

The following year, the Siméon family made their annual migration northward, as recalled by hunter-trapper Gérard Siméon. Gérard was born in the forest in 1922, the same year his grandfather, Malek, passed away. At eighty-four years of age (in 2006), Gérard was among the very last of the hunters of Pointe-Bleue who could recollect bush life before the upheavals brought on by the Second World War. He would live as a hunter-trapper until the 1960s and then as an artisan of Native handicrafts.[12] Gérard passed away in 2009. The 1931–32 migration began in August when a multiple-family group (Siméon, Dominique, Bellefleur, Picard, Benjamin, Natipi, and Raphael) made their way by hired automobiles from the reserve to the lower Peribonka River. Young Gérard would have been nine years old. By canoe, they proceeded upstream to the rapid called Cran Serré. Here they spent three weeks, fishing and snaring hare, and also picking blueberries to sell. Eventually the group continued upstream, the individual families paring off as they reached their respective territories. The Siméon group of six adults and nine children was led by Gérard's father, Thomas, and included two aunts and two adult males from other families. They ascended the Peribonka for around two weeks to a location known as Pakushipashtuk, above the forks of the Manouan. Here they spent the fall and winter trapping seasons, from September through March. The animals trapped included marten, beaver, mink, otter, and muskrat. Hare and various fish supplemented their diet. They descended to the forks in April to spend several weeks trapping muskrat. They also killed and ate a bear. Three fur buyers, Joseph Drolet, Clément Dufour, and Charlot Buckell, met the various Native families at the forks in order to purchase pelts before the trappers reached the trading posts at Lac St-Jean. The families descended together, passing one night at the Raphaels' camp and another at Chutes McLeod, before continuing on to Pointe-Bleue.[13]

In the 1935–36 season, Elie Connolly (then eighteen years of age) joined a large multi-family group departing Pointe-Bleue by truck for the Cran Serré rapid on the lower Peribonka. All told, seventeen adults and ten children made up the party, travelling in thirteen canoes. The group made their way upriver, the St-Onge and Boivin families going up the Rivière Manouan, and the Germain and Connolly families using

the Riviére Serpent/Brodeuse path to skirt Passe Dangereuse. At Lac
Onistagane, the Germains and Connollys went their separate ways, with
the former ascending the Riviére St-Onge and the latter (Elie and his
young wife) proceeding beyond the confluence of the Peribonka-Savane
to the Rivière Nipissi, which borders the territory of the Mistassini
people. Here they spent the autumn hunt, trapping mostly beaver and
mink. The couple returned to Pointe-Bleue to celebrate the winter *fêtes,*
making the lengthy journey by snowshoe and toboggan to the town of
Dolbeau, and then taking a train to the reserve. In March, Elie returned
to the bush for the spring hunt by himself, via train to Dolbeau, horse
and buggy to the edge of town, and then toboggan and snowshoe. This
time, he ascended the Mistassibi River before crossing into the valley
of the Peribonka and ascending to the Nipissi and aux Sables valleys.
Along the way, he snared hare and shot grouse. Elie tended a trapline on
snowshoes for a month or more, then continued trapping and hunting
by canoe after breakup. He met several Mistassini hunters in this vicinity.
In mid-June, Elie loaded his furs and joined the Benjamin family for part
of the journey downstream. They parted company when the Benjamins
headed down the Bersimis River and Elie continued his descent to the
Peribonka-Manouan forks where he met the family of Thomas Siméon.
At Chutes McLeod, he hired a truck to return to Pointe-Bleue.[14]

During the migration of 1936–37, Bersimis hunter Paul Benjamin
took fifty days to reach his family's hunting territory in the Peribonka
tributary, Rivière de la Grande Loutre – an inbound journey of 350
miles. Paul was thirteen and joined his father, mother, and two siblings.
They travelled in two canoes and camped in two canvas tents. Although
the family relied largely on the *petite chasse* of hare, partridge, and fish
during the long and rigorous ascent – "only for our daily needs," as Paul
recalled – they were also lucky enough to kill a moose at the Bersimis's
first portage. It required four weeks to reach Lac Pipmuacan ('Javelin
Lake,' named for an ancient battle between Montagnais and Iroquois
in this vicinity). From its northernmost Baie des Hirondelles (Bay of the
Swallows), the family crossed into the basin of the Rivière Manouan via
Grand Lac Détour. Each of over twenty portages to this point required
six round trips; several trips via canoe were needed to pole the family's
considerable provisions up the numerous rapids. Ten days of additional
upstream work brought them to Lac Manouan, where they crossed
watersheds a second time, via the Rivière Modeste, to the Peribonka
River. Paul recalled a trading outpost in this vicinity that was "held by a
whiteman of Lac St-Jean."

From the Peribonka, the Benjamins ascended the Big Otter River some thirty kilometres to the winter hunting camp. Here, at a site known as Piakunuk the family passed eight months living in log cabins at the main camp and three satellite camps. All told, the territory measured fifteen to twenty square miles. Paul's mother, Marie-Louise, gathered herbs, roots, and other trees "as remedies known to us" and took responsibility for the preparation of food and skins. Paul's father, Jacques, directed the hunting and trapping. To the east was the territory of Joachim and Jean-Baptiste Tsernish of Bersimis.[15] In the course of the winter's hunt, the Benjamins encountered several hunters from the inland Cree post of Mistassini, as well as family groups from Pointe-Bleue. The latter, called Piekukemiuluts ('people of the shallow lake'), included the families of François Germain, Paul Dominique, Noé Germain, Paul Niasipi (Natipi), and Bazo and Charles Boivin.[16] The Benjamins' return journey to Pessamit ('place of the lampreys') was much more rapid than the ascent had been, due to the swollen rivers and to the fact that the family did little hunting along the way. "There were animals of all sorts there," Paul stated. "We could hunt as we needed." In less than two weeks, they had returned to the St Lawrence's estuary and the village. They would remain through mid-August: "waiting until after 15 August (la Fête) to go back up into the woods." And the cycle would begin again.[17]

Adélard Bellefleur partnered with Paul Benjamin as an adult trapper. But in 1938–39, Adélard was but a boy of thirteen, joining his own family at their territory located at the height of land between the Bersimis and Manouan rivers.[18] As this land was closer to the reserve, the Bellefleurs could proceed at a more leisurely pace during the autumn ascent; in fact, they took a full six weeks to reach the winter camp on Lac Pipmuacan's Baie des Hirondelles, two weeks longer than the Benjamins had taken. Young Adélard joined his father, Adélard senior, his mother, Adèle, and two sisters, as well as uncles Sylvestre, Joseph, Pierre, Paul, and their families. The group travelled in eight eighteen-foot wood-canvas canoes purchased at Joseph Miller's store on the reserve. It was Miller, as well, who advanced Adélard senior that year's outfit: each canoe held some 400 pounds of provisions, including tent, stove, blankets, flour (carefully sewn into fifty-pound canvas sacks), and pails of lard. Iron traps had been cached at the winter camp; as only Natives ventured into this region as yet, no one feared for the safety of this equipment. Adélard recalled the nineteen portages below Lac Pipmuacan the longest being a formidable five miles in length; all required five or six round trips to carry "le stock" around the rapids and waterfalls of the lower Bersimis. When the rapids were navigable, the adult men poled the canoes, taking several trips in partially laden crafts for each of the rapids. All of this was rigorous work; but the Bellefleurs did not think of it as such. This was simply a means of feeding themselves; it was "their way of life."

The winter camp, dubbed Plekushi ('mountain like a hat'), sat at the confluence of the Hirondelles and Sylvestre rivers at the terminus of Baie des Hirondelles (a place now flooded due to Hydro-Québec's 1956 Réservoir Pipmuacan). A log cabin, maintained each spring, housed the multi-family group for the duration of the trapping-hunting season. The men worked several circuits in the vicinity for two weeks at a stretch, including along the Rivière Sylvestre, and the Lac Maurice chain to the southwest of the main camp. They travelled by canoe until freeze-up, and thereafter by birch-made snowshoe and toboggan. In the course of their travels, the men visited the St-Onge and Picard families (east of Lac Manouan) as well as the Bacon family (along the Rivière Manouan). They also likely visited one or two of the fur trade outposts now constructed on the large lake, whether that of HBC or of private traders. Marten, mink, otter, and beaver comprised some of the animals trapped. Quite plentiful were the "gros gibier" (big game) of moose (in Innu, 'mush') and woodland caribou ('tuk'). "There was always enough meat," Adélard remembered. Large game was supplemented by small game, as well as pike and trout, which were netted under the ice through the winter months. While the men hunted and trapped, the women were equally busy: they maintained the main camp, split the firewood, snared or shot small game, prepared the furs, dressed the snowshoes with rawhide, and dried fish and meat in order to preserve it. Particularly useful for the men's travels was dried, pulverized meat ('leueikent') that could be mixed with bear fat (or lard) and spread on bannock bread to make a calorie-rich food. Men and women, adults and children alike maintained a deep respect for the animals hunted. Bones must never be placed on the earth, Adélard was taught early on, but instead must be hung in the trees; and no part of the animal was wasted. The black bear was particularly revered. Upon discovery of his winter den, he wasn't shot outright. Rather, the hunter removed the snow at the entrance and called out: "grandfather, come out of your hiding place." Only when the bear had exited the den was he dispatched.

Native animism mingled with Catholicism across the winter hunting season: the Bellefleurs' departure from Bersimis awaited the priest-led August 15 *fête*; Christmas and New Year's feasts marked the end of December; Sundays throughout the winter were set aside as days of rest. In June, the Bellefleurs returned to Bersimis after ten months in the interior; according to tradition, as they rounded the point of the Bersimis River and entered the estuary of the St Lawrence, they fired their guns into the air – both to salute the village and to greet the *curé*.

Charles-Henri Picard, in his early thirties, ascended the Bersimis River with his young wife, Catherine, for the hunting season of 1940–41. Charles noted: "We knew that life would be hard in the woods, but we were so happy to leave for the hunting territory."[19] Like the Benjamins and Bellefleurs, the couple worked their way around the falls and rapids of the lower Bersimis while feeding themselves from the *petite chasse*. They took three weeks to reach Lac Pipmuacan. At this point, they turned up the Bersimis and Pipmuacan rivers, paddling, poling, and portaging another one hundred miles north to the winter camp at Lac Manouanis (Little Manouan Lake), just east of Lac Manouan. This latter section of the route entailed thirty to forty short portages. All told, the journey had taken a month.

At Manouanis, the Picards spent the next eight months hunting, trapping, and fishing on territory that straddled the Manouan and Bersimis watersheds. Charles killed a bear and a moose; he trapped mink, marten, fisher, otter, beaver, and lynx; he caught pike, trout, and carp; in the spring he hunted ducks and geese. Catherine maintained the log cabin, gathered currants and cranberries, and prepared the meat, fish, and skins. She also gave birth to a first child, Elizabeth. Several families of the Pointe-Bleue band visited Charles and Catherine during the course of the winter. Come spring, the Picards took all of a week to descend to the reserve on the swollen Bersimis River. Recalled Charles: "It went fast and we rolled down the river like a car on a road." The speed was welcome, though. "We were out of food," he recalled, and "it was hard."[20]

Technological change is evident in some of these migration stories of the interwar/early war period. At the turn of the century, the bark or canvas canoe was the Innu's sole vehicle of transportation. By the 1920s, Montagnais trappers trading at Lac St-Jean were supplementing the canoe with the use of horse and cart, lake ferry, train, and automobile in order to gain more rapid access to the interior and/or to return more easily to the trading post. The advent of the automobile, in particular, along with a growing rural network of roads around Lac St-Jean, allowed Natives to shorten the annual migration journey by several days. Of course, the new technologies posed a historical trap as well an advantage. Whether hiring a taxi or truck, purchasing and servicing an outboard motor for a canoe (still rare during the interwar years but widespread by the 1950s), or indulging in the purchase of dried beans, peas, rice, apples, or oatmeal to augment other basic foodstuffs (a practice that seems evident by the 1930s[21]), a hunter required the pelts (and thus the cash) to pay for these additional expenses. The continued abundance of game animals was a necessity in this arrangement. Any disturbance to the boreal ecosystem would quickly render the hunter's work unprofitable and impractical. We should note that at Bersimis, the technology of the hunt was not substantially changed: neither road nor railroad had yet penetrated the North Shore, and the reserve's road network did not yet extend beyond the village. At most, hunters bound for the Bersimis River hired a horse and cart to transport provisions two to three kilometres from one of the trading stores (HBC's, Joseph Gagnon's, or Joseph Miller's) to the western edge of the Bersimis estuary at Netaukat[22] and most hunters still portaged their goods the several hundred metres to the quay at the point of land separating the Bersimis and St Lawrence. We are reminded that the phenomenon of the Canadian fur trade looked and acted quite differently according to the varied influence of encroaching civilization, even in regions of close proximity. Although the hunters from Pointe-Bleue and Bersimis shared an extensive and overlapping upland territory, the two groups experienced different intensities of exposure to the white man's technology and economy.

The changes in the fur market evident in these narratives seem to have affected Pointe-Bleue and Bersimis hunters in roughly equal measure. As fur trade historian Arthur Ray has shown, events of the early twentieth century undermined the monopoly of the HBC across the North, altering its long and stable relationship with Natives as both suppliers and customers. Western urbanization drove a surge in fur prices, motivating other traders and firms to compete with the HBC. The First World War's 'black year' of 1914 (when the HBC bought no furs and cut off credit) also opened the way for serious competition, as did the advent of the tractor, the outboard motor, and the bush plane, all of which gave white traders greater access to the interior, either for their own trapping ventures or to reach Native trappers before other traders could.[23] Along the Peribonka, these changes were manifest in the proliferation of traders' outposts, for example, at Lac Manouan, and at the forks, which the HBC called its 'Peribonka Outpost.' For those trading at Bersimis, HBC and private traders opened inland outposts at Lac Pletipi and Lac Pipmuacan to attempt to ensure the Innus' loyalty. No longer did Natives need to make the entire journey to Bersimis or Pointe-Bleue to sell their furs. Cash sales increasingly replaced the barter relationships of the past. Independent traders Joseph Drolet of St-Félicien, Léon Roy of Roberval, and Pierre Doucet of Girardville, as well as Joseph Miller and Joseph Gagnon at Bersimis, offered a variety of prices and removed the incentive to deal exclusively with the HBC. Montagnais-HBC co-operation was "rapidly breaking down," especially among the younger hunters, ethnologist Julius Lips noted in the 1930s.[24]

High fur prices also drew additional trappers into the bush, whether they had ancestral ties to particular terrain or not. The result was increased trapping-hunting pressures on the land (as manifest in David Philippe's narrative) as well as the beginnings of competition among the trappers themselves. While the breakdown of the old, paternalistic order was economically advantageous to the Innu, it was only temporarily so. With the collapse of fur prices in the Great Depression, a number of trappers were forced to seek summer wage labour to supplement their income and procure the tools and necessities of the trapper's trade. Pointe-Bleue Natives served as hunting or fishing guides for tourists, as fire wardens, or as labourers in sawmills. At Bersimis, there was a small amount of work to be done cutting wood, loading wood pulp onto ships, or netting and selling migrating Atlantic salmon. In this way did the Innu's growing dependence on (or intense involvement with) the fur trade in the interwar period act as a segue to wage labour employment and sedentary life in the post-war era.

Despite changes in technology and economy before the Second World War, the life ways of these Montagnais continued much the same. Innus' annual migrations remained long and rigorous, requiring extraordinary bush know-how and skill. Whole families – men, women, and children alike – still made these voyages, the large multi-family departures or returns to the post taking the form of festive annual rituals. Trappers continued to depart the trading posts in August and return in June, generally passing the entire autumn, winter, and spring in the interior. In short, hunting and trapping persisted as the year's predominant activities. Land ownership also stayed collective and flexible, in that primary ownership was vested in the band rather than the individual: through the 1920s and 1930s, unmarried males or young couples might trap a section within another family's territory, with permission of the headman; by this mechanism, over time, territories could change hands. As before, any Native could hunt for food while travelling through the territory of another family, with 'country food' (wild meat and fish) still being the essential and pre-eminent source of Innu life. Rivers, both large and small, remained the essential arteries of that life, providing a means of transportation, orientation, and subsistence. Where watersheds converged at their headwaters, there too did the territories of the Bersimis, Pointe-Bleue, and Mistassini bands. Native-made canoes, paddles, snowshoes, toboggans, and moccasins remained the indispensable means of movement to and about the hunting territories. And the hunter stayed strongly attached to the land: carefully managing the annual use of the territory by section, taking care not to waste meat, offering respect to the animals harvested by the proper disposal of their bones, offering a prayer of thanks at the end of the hunting season. At the summer trading post, Montagnais practised Catholicism and observed the sacraments/rituals of baptism, communion, confirmation, and marriage. But in the bush, Native animism, so well attuned to the practice of the hunt, still held considerable sway.[25]

Certainly, by the Second World War, the region's Innu were increasingly dependent upon the modern tools of the fur trade as well as vulnerable to the market's volatility. At the same time, they remained deeply rooted in the semi-nomadic way of life and still very much dependent upon the land itself. The German ethnologist Lips (having fled Hitler's Germany for the United States in 1934) did field work among the Montagnais-Naskapi in 1935, including those of Lac St-Jean. His "informants" included members of the Kurtness, Kak'wa, Germain, Béjean (Bégin), Connaly (Connolly), Moar, and Jourdain families, as well as the manager of the HBC Pointe-Bleue post, Mr Fawley. Lips found a society of "subarctic hunters" organized according to an age-old system of "family hunting-grounds," notoriously resistant to conversion to agricultural life and still attached to "certain totemistic conceptions" in human-animal relations, including the gifting of animal remains to animal spirits. "Hunting and trapping constitute the backbone of the economic structure of these Indians," wrote Lips.[26] The American anthropologist J. Allan Burgesse spent several months at Pointe-Bleue through the summer season of 1943 to study Native land tenure. In the course of his visit he interviewed hunter-trapper Jack Germain as well as the manager of the HBC post, David Eaton Cooter. Like Lips, Burgesse came away from his visit with the impression of a distinctly bush-oriented Native hunting society. "The Indians of Lac-St-Jean, whose reserve and mission centre are at Pointe-Bleue on the west shore of the lake," wrote Burgesse, "are hunters and gain their livelihood by roaming the forests of the interior, trapping and fishing. They know nothing whatsoever of agriculture."[27] This firm attachment to hunting as an occupation is certainly evident in the remarkable film and photographic record made by Quebec forester Paul Provencher in the 1930s. Provencher's images, recorded among the Natives of the Bersimis Reserve, show a hardy and healthy society, highly skilled in woodcraft and very much immersed in and attuned to life in the boreal forest. While some rare few members of the community dabbled in farming on the reserve during the summer, by tending a potato patch, "it isn't gardening that interests them," Provencher explained. "That isn't a man's work. They are hunters!"[28] And this impression is confirmed by the dozens of interviews conducted with Innu elders in the early 1980s, in both Mashteuiatsh and Pessamit. Collected for treaty negotiations, to document the occupation and use of Aboriginal land, we tapped those same interviews to reconstruct the annual migrations of the pre-war era. Anthropologist Denis Brassard concludes from a synthesis of these interviews at Pointe-Bleue that the annual migration cycle retained a "relative stability" and "regularity" right through the 1930s. Innu hunters of the interwar period still held "a certain autonomy of action" in their lives.[29] Jacques Frenette echoes these conclusions about the hunters of Bersimis: "The life cycle of the Montagnais of Betsiamites during this ... period ... [was] essentially oriented towards the utilization of natural resources. In this way, it [was] undoubtedly true to the traditional way of life of subarctic hunters."[30]

Nor had these hunters yet experienced significant government intervention. The Depression saw the advent of federal relief payments for the sick and destitute in the form of minimal rations at the Hudson's Bay Post (worth $4 to $7 per person per month). This may have begun to draw additional Natives, especially the very old, to live year-round at the posts by the end of the decade.[31] But there was as yet no government dictate tieing family allowances to year-long Aboriginal schooling, a policy that would curtail the generational passage of bush knowledge from parent to child and vastly accelerate sedentary village life after the war. As well, Ottawa had not yet restricted Native hunting in these watersheds, in particular the harvesting of beaver, as Indian Affairs had begun to do in the James Bay region. So, with the exception of the summer visit to the reserve, the Natives were "physically as well as mentally beyond the reach of the white man's legal sphere," Lips asserted of Aboriginal life in the 1930s, "and deep in the woods the Government's Indian legislation ceases to be relevant; our Indians continue to observe the ancient traditions which have governed their lives for centuries."[32] Government's most significant influence up to the Second World War was an indirect one, and based in Quebec City, rather than Ottawa: the provincial government's effort to promote private economic development by the concession of timber limits, mining claims, and hydroelectric resources.

These promotional efforts would soon bring about what amounted to an industrial revolution on the Canadian Shield, with calamitous consequences for Native peoples. Through the pre-industrial era, the Innu held valuable skills in the Canadian fur trade. They were uniquely qualified to harvest animal pelts from the boreal forest – and were also able to sustain themselves from the hard-won bounty of the interior. Through the Second World War, therefore, they retained an autonomy and stability in social, political, and religious life, as we have observed, even while adapting to technological and market changes at large. Of course, this entire macroeconomic arrangement depended upon the continuation of the subarctic fur trade; and the fur trade, in turn, depended upon the availability of huge, remote swaths of boreal land whose paramount economic use, for whites, remained the harvest of fur. When continental and global markets called for pulpwood and minerals and energy, these 'new staples' replaced animal pelts as the white man's most desired items of extraction. Native hunter-trappers lost their value and berth in the marketplace. And the Innu lost their livelihood as skilled artisans of the northern forest.

The industrial revolution of the Peribonka-Bersimis uplands was triggered by provincial resource concessions to Aluminum Company of Canada, a corporation of American origin.

The arrival of Aluminum Company of America at the Saguenay watershed in the 1920s, as well as the resulting creation of the subsidiary Aluminum Company of Canada, Limited, has been noted in the Introduction and described in detail elsewhere.[33] Canadian entrepreneurs first eyed the Saguenay's enormous energy potential for industrial manufacture at the turn of the nineteenth century. However, these projects stalled for want of adequate capital and, in 1914, definitive control of the Saguenay's hydroelectric potential passed into the hands of New York–based tobacco-magnate James B. Duke. His plans for hydro development were delayed by lack of provincial government sanction to store water in Lac St-Jean and then by the economic depression that followed the First World War.[34] Finally, in 1924, Duke attracted the attention of Aluminum Company of America president Arthur Vining Davis who was confronting rising aluminum demand in the United States and saw in the Saguenay's cheap and abundant power the opportunity to retain a North American monopoly of the light metal. Duke and Alcoa merged their interests in 1925, with Alcoa purchasing both a large block of electrical power from Duke's Isle Maligne dam as well as the enormous and as-yet undeveloped waterpowers downstream in the vicinity of the Shipshaw River. Alcoa created a subsidiary called Aluminum Company of Canada, Limited, to hold its Canadian properties. Alcoa also set about constructing an aluminum reduction factory at the new village of Arvida (named after Arthur Vining Davis). Upon Duke's death, A.V. Davis engineered the purchase for Alcoa in 1926 of a controlling interest in the Isle Maligne dam and its crucial water rights. In this way, Alcoa acquired firm territorial control over the production and distribution of the Saguenay's energy, eventually becoming the region's dominant industrial landlord.

Aluminum Company of Canada, Limited gained a measure of independence from its American parent firm in 1928. In that year, motivated by a mixture of corporate strategy and family politics, Alcoa established a corporation called Aluminium Limited, ostensibly as a Canadian firm. A.V. Davis saw a need for a separate managerial team for Alcoa's growing international operations, including reduction plants in Europe and bauxite properties in South America. Equally important were the aspirations of his younger brother, Edward, to command his own company. And so Aluminium Limited was incorporated in Canada; to it were assigned Alcoa's international properties, most important being Aluminum Company of Canada, Limited, in exchange for the stock of the new company, which was distributed to Alcoa's shareholders in the year of its founding. Alcoa executives would remain the majority holders of Aluminium Limited shares (and thus retain control of Aluminum Company of Canada, Limited) until after 1950, when US federal courts ordered that Alcoa shareholders divest themselves of either Alcoa or Aluminium Limited shares, effectively separating the American and Canadian branches of the company. Canadian nationalism accelerated this trend of the 'Canadianization'through the 1960s, including a change in corporate name to Alcan Aluminium Limited. But until then, and certainly through the years of the Second World War, it was well understood in the United States that Aluminium Limited – and its subsidiary Aluminum Company of Canada, Limited – were still closely associated with Alcoa. Within the Roosevelt administration, for example, the Canadian company was dubbed Alcoa's "sister monopoly in Canada" or its "Siamese twin."[35]

Aluminium Limited's US roots included its American management team. As Alcan's first corporate historian (a former Alcoa employee) wrote: Aluminium Limited was directed by "a small nucleus of Alcoa-trained men who volunteered to leave their old company to join the

Canadian organization."[36] And this "small band of corporate warriors" (in the words of Litvak and Maule) led the firm through the 1940s. During the Second World War, when Alcan expanded its Saguenay operations, the firm was domiciled in Canada (its head office in Montreal) and regulated by Canadian law. Nevertheless, its principal executives were American citizens who, like the Boston-born and Harvard-educated Edward Davis, had formed their early careers in the American parent organization.[37] E. Davis, who had worked under his older brother, oversaw Aluminium Limited's operations from his offices in Boston and New York. Ray ('Rip') Powell, vice-president of the Canadian parent firm, Aluminium Limited, and president of Aluminum Company of Canada, Limited during the war, had peddled aluminum cooking utensils in his native Illinois before joining Alcoa's sales staff in 1909. Other key wartime players from the Alcoa organization included Elmer MacDowell, Aluminium Limited's chief sales officer during the war as well as a company director, and McNeely ('Mac') Dubose. Dubose, nicknamed 'king of the Saguenay' by his Alcan colleagues, was an electrical power expert and the resident vice-president of Aluminum Company of Canada, Limited at the Arvida Works. He oversaw the construction of hydroelectric facilities in Quebec (Shipshaw's "directing genius," described *The New York Times*) and later British Columbia. His origins were immediately evident by his North Carolina accent.[38]

Beginning in 1936, demand for the energy-intensive light metal aluminum rose in all those nations that would develop the major industrial war machines of the Second World War. In order of importance for Aluminum Company (in the pre-war period), these included Great Britain, Japan, the United States, and Germany.[39] Of equal importance for the development of the Saguenay watershed, Aluminum Company completed, in the spring of 1937, its purchase from Alcoa of the subsidiary Alcoa Power Company, which held the Chute-à-Caron power station and the undeveloped Shipshaw power site.[40] Henceforth, rising wartime demand for aluminum ingot – to make airplanes – and the availability of large quantities of inexpensive electrical power along the Saguenay River drove the Saguenay basin's exploitation through the war years.

That exploitation might have been restricted to the Saguenay-Lac-St-Jean valley, leaving the boreal uplands alone, were it not for the long memory of the local French-Canadian population. As early as 1915, American engineers and state surveyors understood that it was possible to raise the level of the lake by at least several more feet to substantially increase power production along the Saguenay below.[41] In this sense,

damming the Peribonka's headwaters was hardly inevitable. But local
opposition would surely prevent this, borne as it was of the rough corporate behaviour of 1926 when Duke-Price Power Company, predecessor
to Saguenay Power (an Alcan subsidiary), flooded the lakeshore before
notifying landowners or offering compensation. Certainly this 'tragedy'
stung the local farming population. That Aluminum Company had also
been "staggered" by the resulting property settlements (in the words
of a corporate executive[42]) meant that private enterprise, too, would
proceed with caution in the matter of additional storage in this lake.
So, Aluminum's decision to pursue additional storage higher up in
the watershed was as much an issue of provincial politics as it was of
hydraulic engineering. "Within the limits fixed by the Government,"
explained American consulting engineer Frank Cothran in 1939, "[Lake
St John] has a storage capacity of approximately 150 Billion Cubic Feet.
This amount of storage is quite insufficient for regulation of the natural
annual flow, and is far too small for carrying over any water from a
wet year to succeeding dry ones." By contrast, understood Cothran, the
boreal uplands constituted "a territory that is wholly undeveloped [and]
uninhabited."[43] As Alcan's official historians later put it: it was the "farm
vote" that blocked a further-elevated Lac St-Jean, whereas, "far north
in the uninhabited wilderness ... of the desolate Canadian Shield," a
"remote and largely useless semi-tundra would be flooded."[44]

Of course, as we have seen, this terrain was neither "uninhabited,"
"desolate," or "useless" for those who made their living as hunters and
trappers of the northern forest. The simple fact that farmers voted and
Aboriginals did not – nor did Natives hold any meaningful voice in
Quebec society – explains in large measure the present location of lakes
Manouan and Passe Dangereuse.

From the summer of 1937, officials of Aluminum's subsidiary
Saguenay Power Company sought additional water-storage capacity
for its Isle Maligne and Chute-à-Caron power plants. A survey team
directed by engineer Fred Lawton and guided by local Innu, including Jack Germain, laboured from late August into October 1937 along
the upper Manouan River. Lawton's team returned to recommend the
construction of a stone-filled timber crib dam at the headwaters of the
Manouan River near the same location where government surveyors
had recommended a storage dam in the previous decade. Aluminum's
dam would be but thirty feet high and raise the existing lake chain in
this vicinity a modest sixteen feet. Nevertheless, the dam would create a
substantial reservoir (roughly one-third the size of either Lac Saint-Jean

or the Gouin reservoir) by enlarging the existing lake chain by over fifty square miles, capturing the flow of some 1,800 square miles of territory, and augmenting the Peribonka's winter low-water flow by 5,500 cubic feet per second. Moreover, this extra flow volume could be achieved at the reasonable expense of under half a million dollars.[45]

Saguenay Power general superintendent Dubose submitted his formal request to create such a reservoir to Quebec's Department of Lands and Forests in November 1937. Directing his request to the acting chief of the Hydraulic Service, A.B. Normandin, Dubose promised "big help to labor" with this Depression-era construction project, and he urged "lowest possible charge by the Government" for the rights to flood the Crown's timbered land and utilize the additional regulated flow of a provincial waterway to generate electricity.[46] We should note that the plans submitted to the Hydraulic Service chief used maps and benchmarks prepared by the provincial government's Quebec Streams Commission earlier in the decade; thus the state's preparatory investigations of the region served private industry and industrial development as they were meant to do. Dubose had also, by this point, laid some political groundwork, having spoken in person with Avila Bédard, deputy minister of lands and forests, as well as the Conservative/Union Nationale provincial representative of Lac St-Jean County, Joseph-Léonard Duguay. Duguay, in turn, had mentioned Aluminum's interest in a concession to Premier Maurice Duplessis, who also held the Lands and Forests dossier during the first year of his administration.[47]

This initial request, however, proved to be a false start to the Lac Manouan negotiations. By the early weeks of 1938, the recession of 1937–38 had gripped North American markets, a fact evident to Aluminum in the drop in power demand by the paper mills they served, especially newsprint producer Price Brothers. Anticipated power shortfalls didn't materialize, including the 100,000 horsepower shortage that justified Lac Manouan. Aluminum was once again, as in the early Depression years, struggling to sell its surplus electrical power. In January 1938, Dubose informed the Hydraulic Service chief Normandin that, for the time being at least, the Lac Manouan project had been put on hold.[48]

Within the year, it was on again; but this time, it would be the provincial government that backed out. Late in 1938, Dubose returned to the provincial government for a concession at Lac Manouan, again directing his request to the chief of the Hydraulic Service, Normandin.[49] Curiously, the record shows there was no response from Normandin. Instead, following an unexplained delay of several months, we find evidence that in May 1939 Aluminum Company applied directly to Premier Duplessis. We should note that Duplessis no longer served as minister of lands and forests, having handed that portfolio to his former minister of public works, John Bourque. Duplessis then directed civil servant Normandin to advise him on a decision. Thus it was that the provincial premier took charge of a bureaucratic process delegated, according to provincial law, to the technical experts of a Quebec ministry. A change was also evident in the terms of the new draft accord: it made no mention of the standard annual power tax (in this era, 50 cents per horsepower/ year) to be charged Saguenay Power for the additional energy generated downstream as a result of the regulated flow of water. Instead, Aluminum Company was to pay the government a flat fee of $50,000 for all the properties and rights conceded – an outright purchase rather than a lease of the Crown's lands and rights in the bed and banks of Lac Manouan, a practice out of date in Quebec for several decades.[50]

Not surprisingly, Normandin recommended against a grant in this form. He noted that the concession, as written, failed to assure a continuous flow of water in the lower Peribonka River, putting in jeopardy future private or public hydroelectric developments of the Peribonka's Chute-à-la-Savane and Chute-du-Diable. He pointed out that Aluminum Company still held some 150,000 surplus horsepower for industrial production that was available at its Chute-à-Caron plant. He made clear that once this power was absorbed by industry, and the Lac Manouan project undertaken, the government stood to gain far more revenue if a grant was executed as a long-term lease rather than sale. Finally, he challenged Aluminum Company's assertion that the thirty square miles or so of timber to be flooded had little to no marketable value, referring the matter of the wood's valuation to the experts of Quebec's Forest Service. Normandin recommended a long-term lease (the typical form of grant since roughly 1910) rather than a sale, and suggested financial terms that would take into account the remoteness and difficulty of industrial development of the site ($10,000–15,000/year, the amount to increase to $25,000/year after the first fifteen years).[51]

Clearly, Normandin should have been involved in crafting the accord in the first place. Bringing in the waterpower expert at this late stage of negotiation, with an Order-in-Council already in draft form, was certainly inefficient, if not inept, on the part of the Duplessis administration. That Normandin felt free to point out the accord's limitations suggests something of his integrity as a civil servant – and/or his unusual sense of job security under a politician known for his irascible and authoritarian style.[52] In fact, Duplessis seems to have followed Normandin's advice. The record tells us simply that Aluminum's request of May 1939 was "refused" by the premier.[53]

Why did Duplessis, well known by contemporaries and historians alike as a "protector of big corporations" and "great friend of private enterprise,"[54] reject Aluminum Company's request in the matter of Lac Manouan? As was quite typical, the premier left no evidence with which to interpret his decision. We can only speculate, and the most probable explanation involves politics. The control of the electrical power industry by Anglo-American capital was among the most volatile political issues of the interwar period in Quebec, second only to federal-provincial relations. Duplessis came to power in 1936, committed to nationalize the industry and rein in the giant power 'trust' (with whom, he claimed, the former Liberal administration of Louis-Alexandre Taschereau had been excessively cordial and permissive). Yet the premier had not acted in this regard, having barred from cabinet posts several key nationalists and/or reformers determined to check the influence of private enterprise in the power industry and, shortly thereafter, seeing the defection of others.[55] What small legislative actions the Union Nationale had proposed to increase state control over the hydro industry – to establish the next permutation of advisory board for electricity rates or to build a small power plant in the Abitibi region with public funds – were correctly deemed, by nationalist critics, as but lip service. And in May 1939, Duplessis would soon be facing his first re-election campaign.[56]

In short, rejecting Aluminum's demands in the spring of 1939 might provide some additional political camouflage for the premier's fundamental inaction on the power front during his first administration, something of a tangible record of accomplishment as a trustbuster, as well as a basis for anti-trust rhetoric. Interestingly, we find no evidence that the premier exploited his decision regarding Aluminum Company during the 1939 campaign. It may well have been that he felt no need to do so, being fully confident in an election victory.[57] Only later, having lost power to the Liberals, did the now-leader of the Opposition repeatedly cite his refusal to Aluminum Company as evidence of his nationalism, of his defense of provincial hydro resources against the aggressive encroachments by foreign capitalists. "I rejected [the] request," Duplessis would boast repeatedly in the Assembly, "saying that never would the province of Quebec become the province of aluminum."[58]

Another possible reason for Duplessis's rejection of Alcan's request was that the Second World War had not yet forced the issue of power scarcity for war industries. As of 1939, Aluminum Company was still burdened by 150,000 unused horsepower: "considered surplus and dumped on the market at anything we could get for it," recalled company president Ray Powell. Then, with the war's outbreak in September, came urgent requests from Britain's Ministry of Supply for more metal. By this point the surplus had been reduced by half; but these new requests promised to absorb the Saguenay's remaining power surplus and more power still, perhaps 100,000 horsepower.[59] Aluminum's chief salesman MacDowell flew to London to make the contractual arrangements to sell the ingot. By early October, the first of several contracts was agreed to. By early February 1940, Aluminum had formalized arrangements whereby the company would comprehensively expand its aluminum production facilities by drawing on loans from the British government. In all, demand from Great Britain roughly doubled the company's ingot production capacity early in the war (from a pre-war figure of 90,000 tons to 200,000 tons annually).[60] From the jungles of British Guyana to the boreal uplands of the Saguenay watershed, the British war effort drove Aluminum Company's phenomenal growth in the early war years: mining of bauxite (in Guyana) and fluorspar (in Newfoundland), constructing of port facilities (in Trinidad), expanding of factories for smelting (at the Saguenay) and fabricating (in Ontario and the UK), and, high in the valley of the Peribonka River, erecting of dams to store water and produce electricity – and with it all came attendant changes in global landscapes. So it was that Lac Manouan was created on one of the war's subarctic fronts to amass materiel for the Battle of Britain.[61]

The newly installed government of Adélard Godbout would prove a responsive and co-operative party to such change in the province of Quebec. Though, as in the case of Premier Duplessis, there is good reason to ask why. As a Liberal, Godbout favoured market-driven industrialization as a means of uplifting the economic fortunes of French Canadians. Yet he was also a true believer in democratic process. He refused to favour corporate interests at the expense of others (unlike his Liberal predecessor L.A. Taschereau, the architect of the 'la tragédie du Lac Saint-Jean') and tended to delegate decisions to ministers and civil servants, rather than retaining or hoarding political control for himself (as did both Taschereau and Duplessis). In fact, Godbout was as much social progressive as Liberal, at least in the sense that Godbout the agronomist would leave behind an important progressive legacy, including legislation to nationalize hydroelectricity, create universal education, encourage scientific agriculture, press forward rural electrification, limit government patronage, and grant Quebec women the right to vote.[62] And no progressive would easily cede a watershed of enormous hydroelectric potential to private corporate development.

Neither liberalism nor progressivism explains the Godbout government's Lac Manouan concession the way the war does. The new premier took office on 8 November 1939, having beaten Duplessis and the Union Nationale in a provincial election focused on Canada's, and especially Quebec's, participation in the war. Duplessis opposed participation on the grounds of provincial autonomy and possible conscription of French Canadians; Godbout, a strong believer in Canadian unity, remained open to collaboration with Ottawa in the service of defeating fascism and Hitler. The war was enough to change the direction of resource policy in the matter of the Saguenay's further hydroelectric development. It was Godbout's willingness to fight the Second World War that best explains his relationship to Aluminum Company of Canada during the 1939–44 period of his administration, beginning with the concession of Lac Manouan. As Godbout's close colleague and the new minister of lands and forests, Pierre-Émile Côté explained the Lac Manouan decision during the war years: "The war provided to the government an argument that didn't exist [previously] ... We need electricity for the war effort."[63]

Within days of Godbout taking power (and the UK ingot contract now pending), Aluminum Company pursued a concession in Quebec City, in what would be the corporation's third and final effort to win permission to construct a storage reservoir on the upper Manouan River. On November 10, Aluminum's long-time attorney and member of the board of directors Aimé Geoffrion, on behalf of the subsidiary Saguenay Power, broached lands and forests minister Côté with the now "urgent" matter of Lac Manouan.[64] Côté, for his part, brought Hydraulic Service chief A.B. Normandin into the talks and expressed his willingness to negotiate a deal. The negotiations in late November involved not only Côté and Normandin but also deputy minister Bédard and, on the company's side, superintendent of Saguenay Power, electrical engineer Dubose. Normandin and Dubose, as the technical experts, took the lead in these talks and then brought their results and recommendations to their respective superiors.

Normandin, like his Liberal mentor at the Hydraulic Service, Arthur Amos,[65] would use these contract negotiations to attempt to garner from the development of the province's waterpowers a good measure of both industrial development and provincial revenue,[66] although the war's urgency made these longer term objectives difficult to achieve. Normandin felt strongly that the government should reserve the right to control the Peribonka's flow, given the hydroelectric potential of the lower falls of the river. And since he assumed, with company officials, that Lac Manouan's storage would only be useful during the war years – producing power surpluses thereafter – he pressed forward the idea that the lease be a shorter term than was usual (perhaps thirty years rather than the customary forty or fifty), even reserving to the government the right to revoke Aluminum's control of Lac Manouan directly after the war ended. (That Aluminum was prepared to accept such a fragile agreement is testament to the fact that company officials shared Normandin's sense that this was a wartime project only. Neither government nor company, after all, could anticipate burgeoning consumer and strategic demand for aluminum in the post-war period.) Normandin was equally aware of the dire wartime need to supply Britain with aluminum ingot, as well as the technical difficulty of the construction project, whose cost had now grown to an estimated $700,000, in such a remote location. So he continued to recommend a very reasonable annual rent. In fact, in lieu of the traditional power tax (based on horsepower accruing from the stored water, usually 50 cents per horsepower paid by private industry), Normandin was prepared to recommend the $10,000–$15,000 per year rental fees on the Crown lands to be submerged, in addition to a one-time lump-sum payment for the destroyed timber, which Bédard put at $30,000. In any case, Normandin was aware that whatever power accrued from the additional water might be at least partially taxed as it flowed through the turbines of the Isle Maligne dam on the Saguenay, where Aluminum was paying 50 cents per horsepower for the water stored in Lac St-Jean (adding perhaps $10,000 in revenue annually to some $60,000 a year).[67] Although he would continue to argue for an even-lower rental fee, Saguenay Power's Dubose was prepared to recommend such an agreement to his principals in Montreal.[68]

Once Aluminum consummated its contract with the British Ministry of Supply on 2 February 1940, the firm returned to the provincial bargaining table to complete its accord with Quebec. On 8 February, president R.E. Powell enjoyed an audience with Premier Godbout, minister of lands and forests Côté, and others of the provincial cabinet. Before this distinguished group, the aluminum man explained the urgent needs of the war and laid out his company's plans to expand ingot production in the province, including the plan to store water in Lac Manouan. Precisely how Godbout or his ministers responded to

Powell's pitch is lost to the record; according to Powell himself, the cabinet agreed to "permit and encourage" the building of the dam and, in fact, "approved" the plan as it was outlined at this meeting: a 1,250-foot-long dam at the outlet of Lac Manouan, creating a reservoir roughly one-third the area of Lac St-Jean, augmenting the year-round flow of the Saguenay River by some 3,000 cubic feet per second.[69]

Formal approval of a concession, Powell knew well, could come only from the lieutenant-governor in Council following formal application (with plans and specifications) to the Department of Lands and Forests and, in turn, a recommendation by this ministry to the cabinet. The law was clear in this regard, and had been so since 1918.[70] Complete plans and specifications had not yet been submitted and approved; public notice had not been given; nor were the terms of a concession worked out and agreed to as they had to be since Lac Manouan's creation would inundate a broad expanse of public land. Yet if Godbout himself was supportive of the project, perhaps, thought Powell, Aluminum Company could commence dam construction before first obtaining legal authorization, and thus accelerate the wartime project. On 6 April, Powell wrote Godbout that "the dam at Lake Manouan ... should be begun without much more delay," asking "if you will be good enough to cause to be issued some form of authority ... and thus put me in a position to cause the construction to be started."[71] Godbout did not reply to Powell's request. Evidently the premier was unwilling in this case to waive or bend provincial law, even if that meant serving the war effort. Aluminum Company would not commence dam construction until August when the approval process was completed.[72]

Still, Powell was correct in understanding that the government supported the plan to build a reservoir; the remaining talks focused on the financials, and never on whether Aluminum Company would be allowed to go forward with the project. Powell would continue to press the government for a lower annual rent, arguing that the dam was a matter of wartime need only, there being no assurance of a post-war market for the additional power. Powell pressed as well for the addition of a clause in the contract that would exempt the company from payment if or when the reservoir was not in use.[73] Normandin and his superiors were flexible on the first point – ultimately lowering the rent to $8,000 annually ($25,000 after the first twenty years) and the timber damage payment to $25,000 – but rigidly opposed to the second. Aluminum would pay its modest annual fee for use of the Crown's land regardless of whether Lac Manouan's storage capacity was useful to the company or not. However, Aluminum seems to have won something of an escape clause from this arrangement. By mid-summer, not only was the lease period set at twenty years, renewable for an additional ten (so, thirty years in total), and the Crown able to reassert control of the river's flow anytime before this (as Normandin had insisted), but significantly, Aluminum held the right to abandon the lease "a year after the cessation of the war" as long as the company left all lands, rights, and properties (including the dam itself) to the Crown. Here is further evidence that in 1940 both the government and the company considered Lac Manouan as a matter of temporary wartime need, quite likely to be superfluous when the war was over.[74]

Within several years, Duplessis would publicly accuse the Godbout government of having given away the upper Manouan "for a song," a well-worn phrase used by nationalist critics of Liberal administration resource concessions across the decades.[75] Duplessis, of course, sought to gain political traction by his criticism, which was levelled without elaboration or evidence. The "song" Duplessis referred to was a lease, rather than a sale agreement, and certainly of greater value to Quebec than the fire-sale purchase prices for waters or waterpowers paid to the government at the turn of the century. And in the summer of 1940, with Western Europe overrun by the Nazis and Britain fighting for its life in the skies, there was the war to consider. From Normandin's (or Côté's or Godbout's) perspective, this was a project of wartime necessity. The contract terms fit what seemed at the time to be an unusual and temporary spike in power demand brought on by the war; and the financials took account of the remoteness of the construction site and the elevated wartime costs of labour.

Yet there is merit to Duplessis's accusation. As Normandin's successor at the Hydraulic Service, Raymond Latreille, would point out: failing to tax the power that could be produced with the additional water storage cost the government thousands of dollars of revenue annually, perhaps $19,000 a year for the first twenty years of the contract. Moreover, the vast majority of reservoirs in Quebec were under state control, whether they were constructed by the state or by private industry, whereas Normandin's terms gave Aluminum Company at least thirty years of ownership and use. As the state itself was aware in conceding Lac Manouan: "the Government departs from the long-established policy governing similar cases, owing to the difficulties inherent to the realization of this project in a remote country and also taking into account that the proposed storage works are, in part, related to the fulfillment of … important war contracts."[76]

Latreille would later put it more simply: the agreement regarding Lac Manouan constituted "favourable treatment" to a private firm.[77]

Various formalities were fulfilled by summer's end. Aluminum Company prepared and then submitted the requisite plans and specifications of the dam and reservoir to the Department of Lands and Forests as well as the local registry offices in Roberval, Chicoutimi, and Hébertville. The government inserted requirements in the contract to construct a fish ladder and log-slide, lest sport fishermen or lumbermen reach the remote latitudes of the upper Manouan River in the coming years. Drafts of the future Order-in-Council were subtly revised by Aluminum's attorneys Geoffrion-Prud'homme and by Quebec's civil servants.[78] And there was a change in personnel inside the Hydraulic Service: after twenty-eight years with the Hydraulic Service, Normandin departed to serve on the provincial Public Service Board and senior civil engineer Latreille, a future founding commissioner of Hydro-Québec, took up the role as chief.[79] But the contract's basic terms, already worked out in the early months of 1940, went otherwise unchanged. Late in July, Aluminum Company informed the government once again of its intention to proceed with the construction of the storage dam, the company's construction schedule requiring completion in time to capture the spring flood of 1941. On 7 August 1940, with the company urging the government to issue authorization, the Godbout government passed the Order-in-Council approving the creation of Lac Manouan.[80] A political process begun in 1937 under the previous regime came to a close.

With the bureaucratic-political hurdle cleared, construction could commence. Lacking the time to build an access road of over one hundred miles into the bush, Aluminum Company began a massive airlift of all necessary workers, draft animals, and equipment in order to build a 1,250-foot-long rock and timber dam in subarctic latitudes during the severe months of winter. By early April 1941, the gates of the dam were closed and storage of water had begun. By the following autumn, the reservoir was filled to capacity, ready to release its water for power production across the low-water period.[81] The achievement of Lac Manouan's rapid impoundment (the result of "the largest air freight contract in Canadian history") is celebrated in several company memoirs.[82] The accomplishment also speaks to the streamlined efficiency of private enterprise, compared to the relative sluggishness and complexity of business-government interaction: the political negotiations had spanned four years and two administrations; actual dam construction took but ten months.

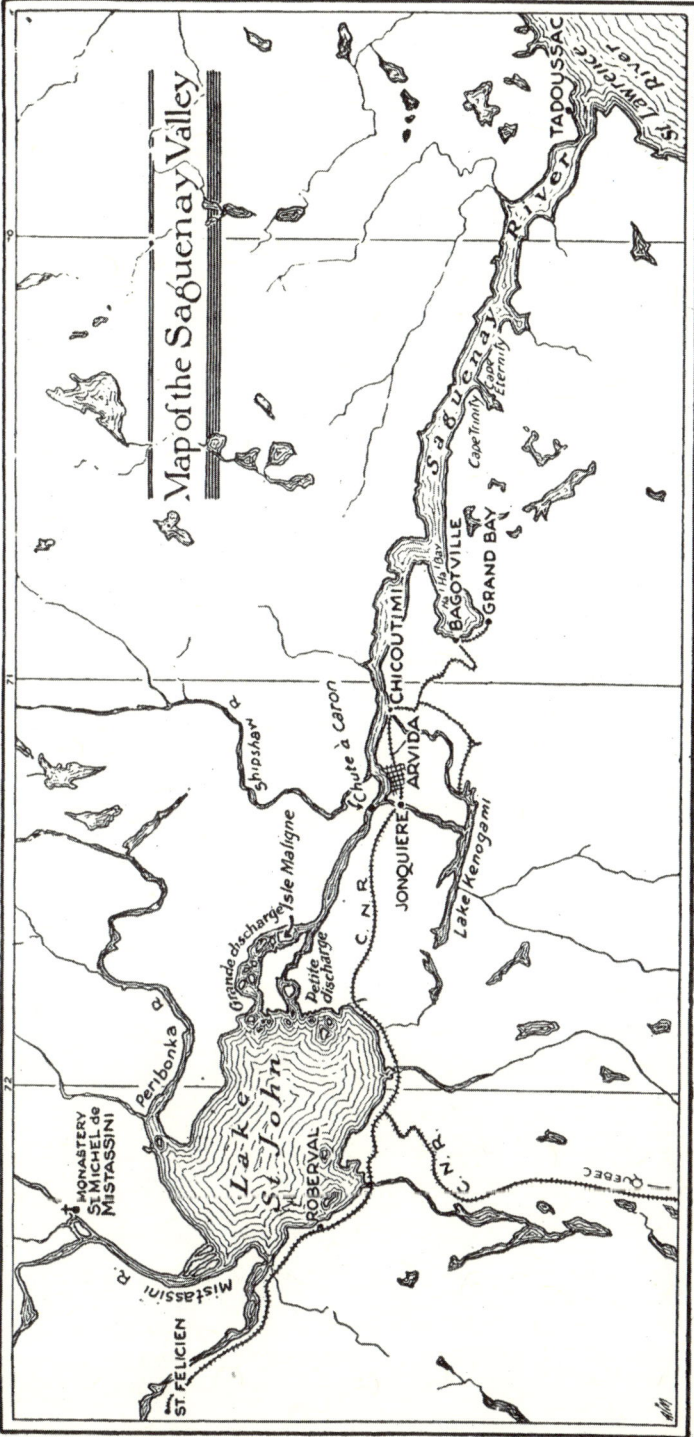

Map of the Saguenay Valley

Saguenay Valley under development. Large-scale hydroelectric development was well underway in the Saguenay Valley when science writer E.E. Free published this attractive map in 1926. The Duke-Price Power Company had just completed its massive power station at Isle Maligne near the outlet of Lake St John to tap the storage capacity of the impounded lake. Further downstream, Aluminum Company of America (parent to Aluminum Company of Canada) had built ingot smelters and the company town of Arvida. The US corporation would soon begin construction of an additional power plant at Chute-à-Caron. (From E.E. Free, "The New Empire of the Saguenay," *American Review of Reviews* 74, 4 [October 1926]: 385–96).

PLANCHE XLIX

LA COMMISSION DES EAUX COURANTES DE QUEBEC

RIVIÈRE PÉRIBONCA
PROFIL EN LONG ET POINTS DE REPÈRE
DEPUIS HONFLEUR JUSQU'À CHUTE MCLEOD

Montréal, Octobre 1920

Ingénieur en Chef

REFERENCES .... Carnet No 267

Ce plan a été publié comme planche XLI du rapport de 1920
avant les révisions faites en 1927 et 1930.

Pour réduire au niveau moyen de la mer
il faut ajouter 215.? d'après BM G.S.C. 941 Bat.335 ont
(Voir publication No8 page de Geodetic Survey of Canada
et plan C.F.C. Z-2500 Filiation between datums)

Note.— Les hauteurs sont données par rapport au Nivellic. *Renvoi* le 23 mars 1:77 pour la partie en aval de Honfleur
Point de départ sur le zéro de l'échelle
hydrométrique installée sur
le quai à Honfleur — 0 = 103.46(C.E.C Datum.)

Profil de la rivière le 6 mars 1921
— — le 28 avril . . 5° de l'échelle à Honfleur. El. 105.14. (.MS.-0.173)
· · · · le 8 Janvier 1930 . pour l'eau haute extrême de printemps de 1928

El. 105.14 .....
El. 105.14 ....
El. 105.14....

Eau à Péribonka El. 107.85 . 7/3/91
D'après Echelle du B. Power Co.

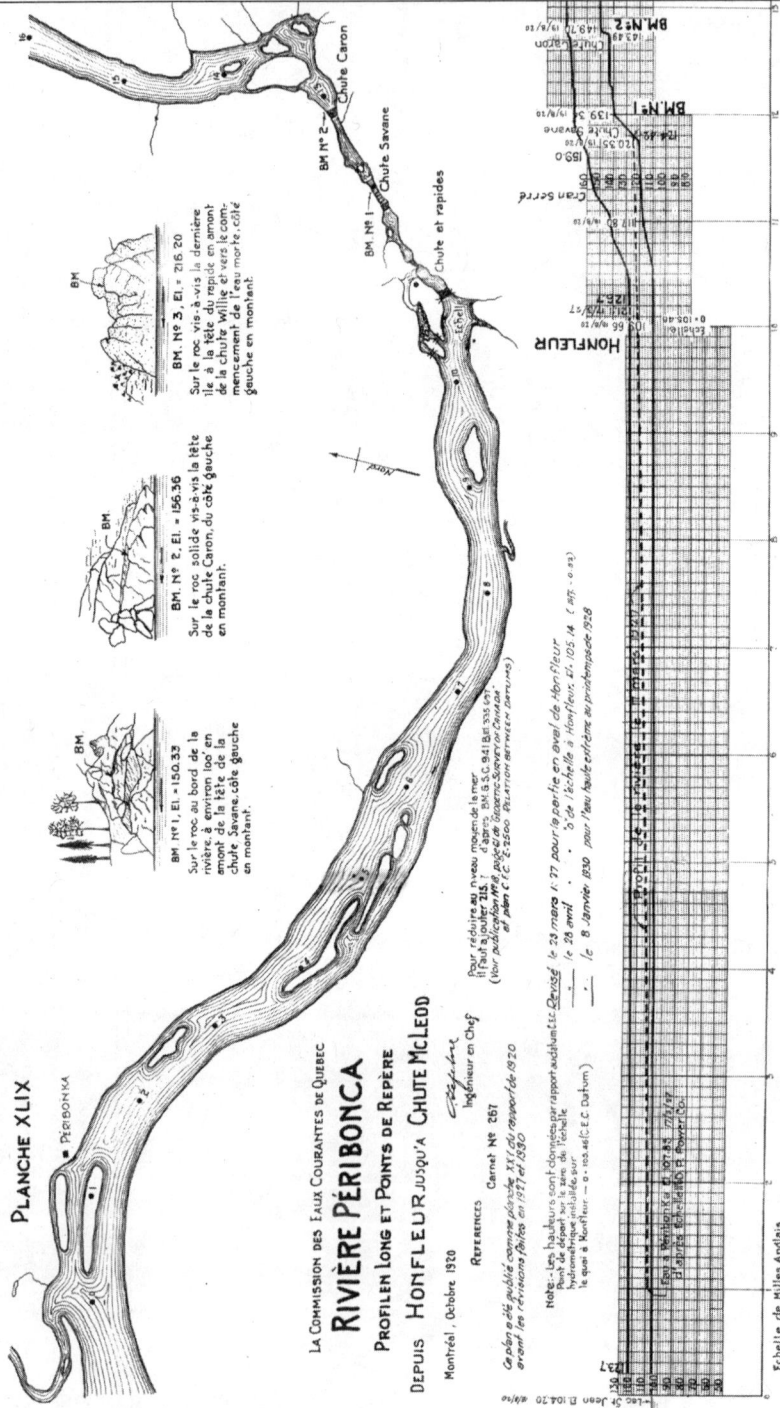

BM No 1, El. = 150.33
Sur le roc au bord de la
rivière, à environ 100' en
amont de la tête de la
chute Savane, côté gauche
en montant.

BM No 2, El. = 156.36
Sur le roc solide vis-à-vis la tête
de la chute Caron, du côté gauche
en montant.

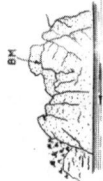

BM No 3, El. = 216.20
Sur le roc, vis-à-vis la dernière
île à la tête du rapide en amont
de la chute Willie, et vers le com-
mencement de l'eau morte, côté
gauche en montant.

PERIBONKA

Chute Caron
Chute Savane
BM No 2
BM No 1
Chute et rapides

Nord

Echelle de Milles Anglais

HONFLEUR
Cran Serré
Chute Savane
Chute Caron
BM No 1
BM No 2

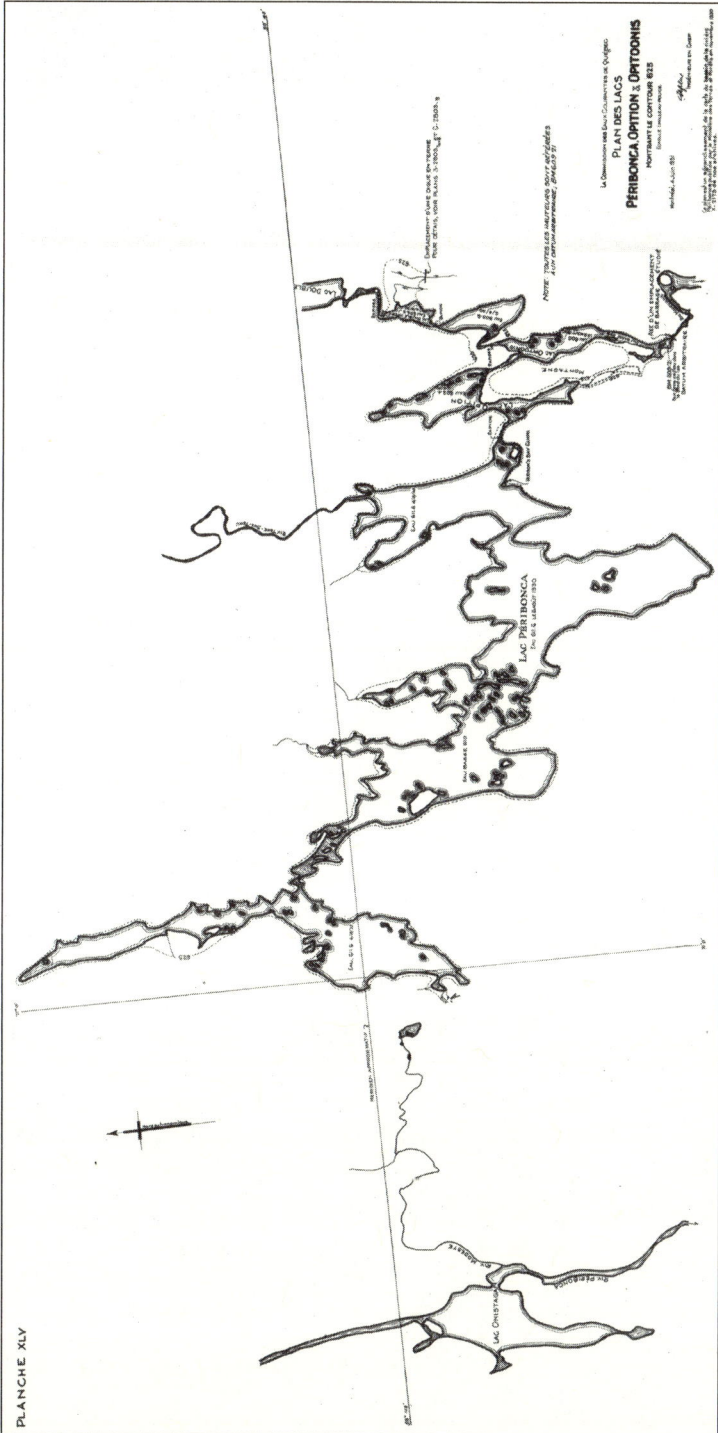

The Quebec state looks upstream. American demands for broader, watershed-wide control of the Saguenay prompted the Quebec Government's Streams Commission to study the Saguenay's major tributary, the Peribonka River. Starting at the river's mouth, surveyors carefully profiled the Peribonka, establishing benchmarks and noting the major falls and rapids, as seen in this detail from a 1920 report. By the end of the decade, with aid of the airplane, the Streams Commission had reached the headwaters of the Peribonka's tributary, the Rivière Manouan. Here, Commission engineers envisioned a large storage reservoir that would link and enlarge an existing chain of lakes. Such projects were delayed by the Depression. (Map 4: *Dix-Neuvième Rapport de la Commission des eaux courantes*, 1930; Map 5: *Vingtième Rapport de la Commission des Eaux Courantes*, 1931)

ALUMINUM REDUCTION PLANTS, U.S. AND CANADA, AS OF 1944
(ANNUAL CAPACITIES IN MILLIONS OF POUNDS)

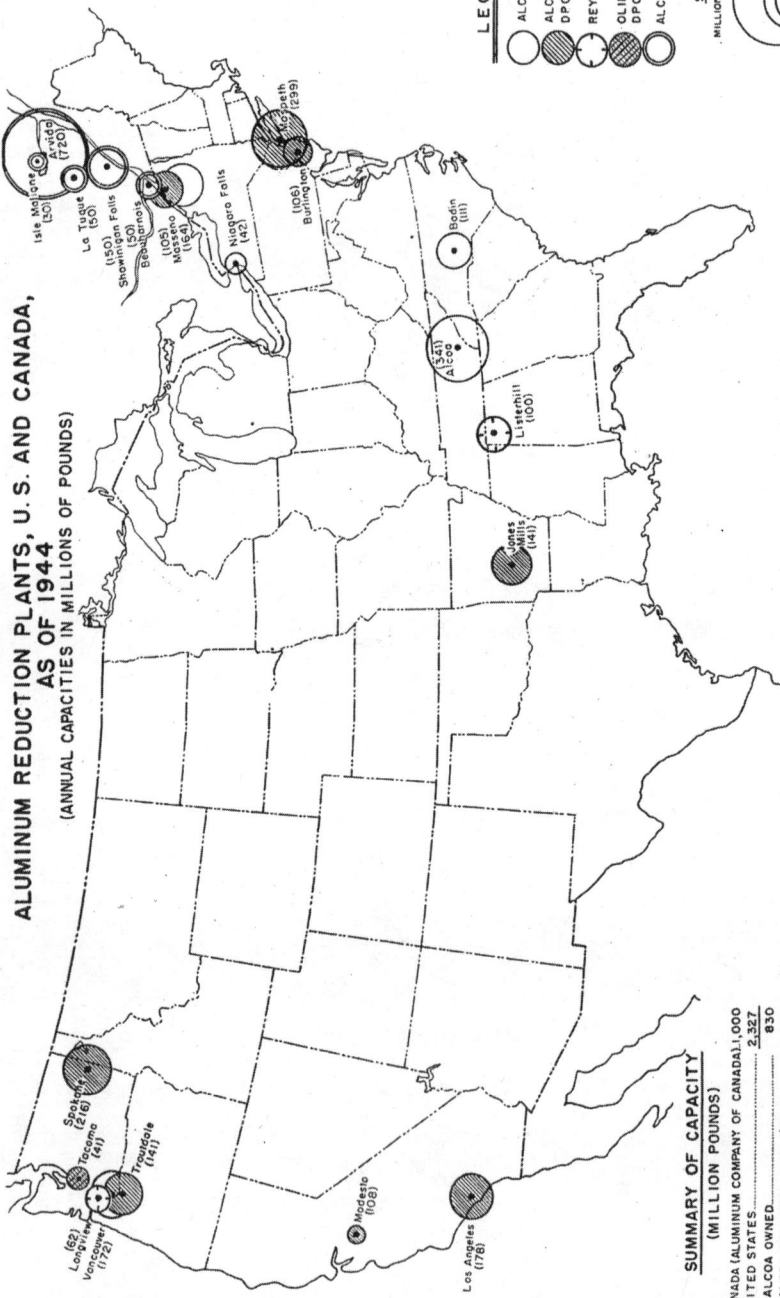

LEGEND

- ALCOA OWNED
- ALCOA OPERATED, DPC OWNED
- REYNOLDS OWNED
- OLIN OPERATED, DPC OWNED
- ALCOCAN OWNED

SCALE
MILLIONS OF POUNDS
720
288
180
108
50
15
0

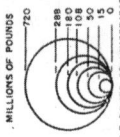

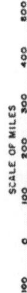

DRAWN BY: C.O.D. BONNEVILLE POWER ADMINISTRATION

Isle Malgne (720)
Arvida (20)
La Tuque (50)
Shawinigan Falls (50)
Beauharnois (50)
Massena (164)
Niagara Falls (42)
Burlington (106)
Nospeth (299)
Bodin (111)
Alcoa (341)
Listerhill (100)
Jones Mills (141)

Spokane (216)
Tacoma (41)
Troutdale (141)
Longview (62)
Vancouver (172)
Modesto (108)
Los Angeles (176)

SCALE OF MILES
100   0   100   200   300   400   500

SUMMARY OF CAPACITY
(MILLION POUNDS)

ANADA (ALUMINUM COMPANY OF CANADA)	1,000
NITED STATES	2,327
ALCOA OWNED	830
ALCOA OPERATED, DPC OWNED	1,294
REYNOLDS OWNED	162
OLIN OPERATED, DPC OWNED	41

SOURCE: WAR PRODUCTION BOARD AND BUREAU OF MINES

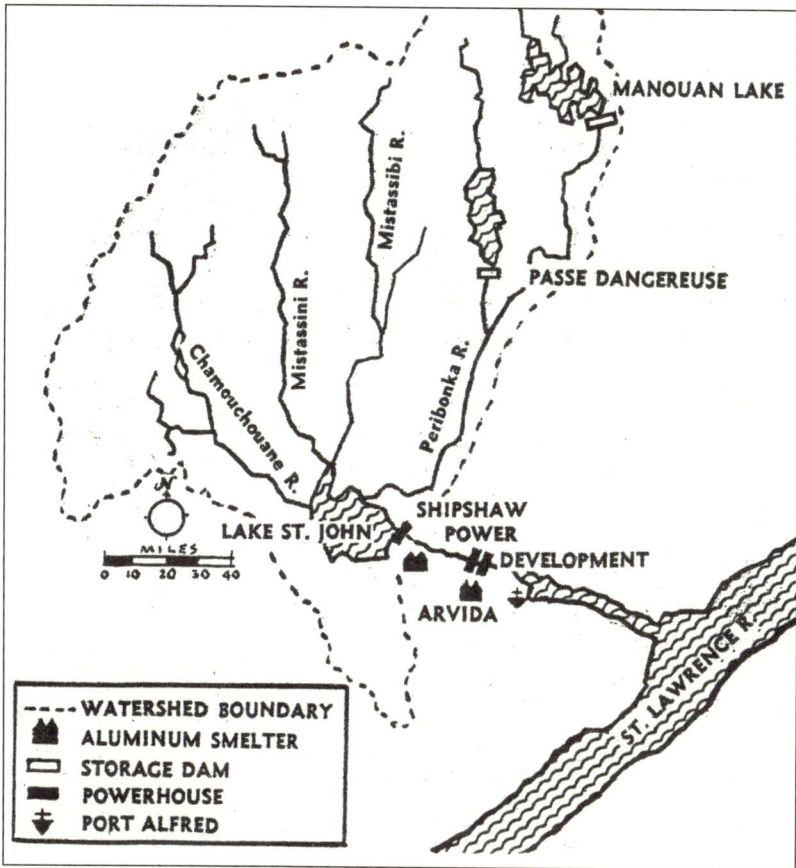

The war prompts hydroelectric expansion. By 1944, Aluminum Company
of Canada's hydroelectric works tapped the entire Saguenay watershed. The
Chute-à-Caron plant had been replaced by the enormous Shipshaw Power
Development. In addition to Lake St John, the company had impounded
and controlled two additional storage reservoirs at Manouan Lake and Passe
Dangereuse. These reservoirs regulated the flow of the Peribonka River
for power production and industrial manufacture well downstream. (Map
adapted from 1956 news clipping, Rio Tinto Alcan, Montreal, 0002619)

OPPOSITE: Canada on the US map of war production. The Second World
War, history's first air war, triggered rapid expansion in aluminum production
by the industrial nations involved. With Aluminum Company of Canada's
construction of the Shipshaw dam, Canada became a major supplier of ingot
to the United States. Aluminum Company of America's operations at Massena,
near the St Lawrence River, reflect both public and private ownership.
Aluminum Company of Canada's private operations at the Saguenay
constituted by far the largest aluminum reduction effort on the continent.
(US Department of the Interior/Bonneville Power Administration, 1944,
collection War Production Board, file 523.421)

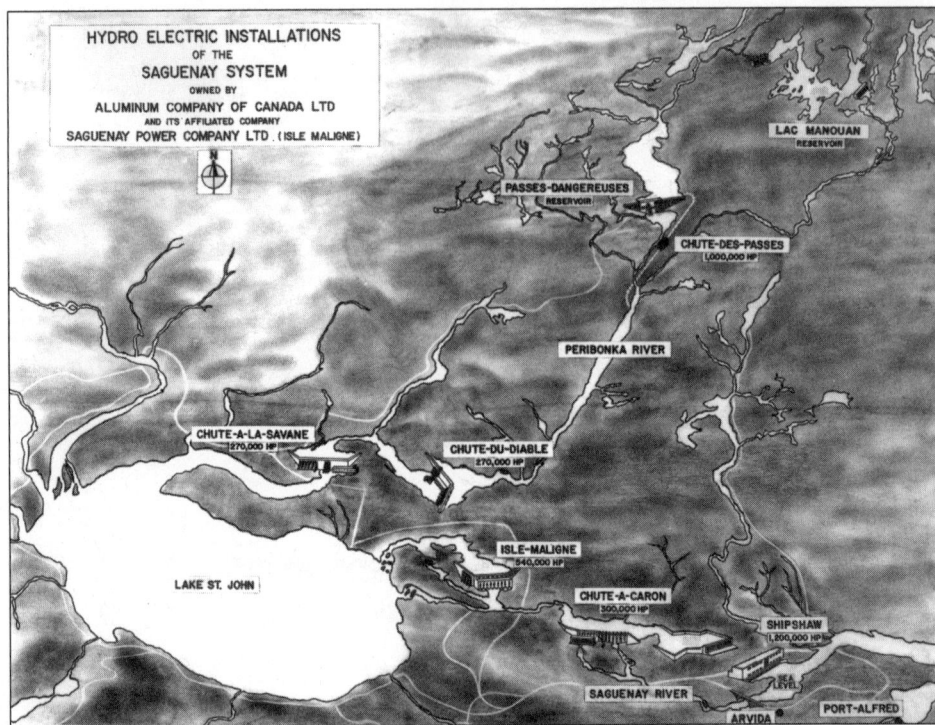

HYDRO ELECTRIC INSTALLATIONS
OF THE
SAGUENAY SYSTEM
OWNED BY
ALUMINUM COMPANY OF CANADA LTD
AND ITS AFFILIATED COMPANY
SAGUENAY POWER COMPANY LTD . (ISLE MALIGNE)

Hydroelectric installations of the Saguenay system. By 1960, Aluminum Company of Canada had added substantially to its power facilities in the watershed, as demonstrated in the company's schematic drawing of that year. Three new power plants now checked the flow of the Peribonka: at Chute-du-Diable and Chute-à-la-Savane (1953), as well as Chute-des-Passes (1960). The corporation's vision of utilitarian control of the drainage basin is evident in this image, as it is in another corporate map of this same era (publication prohibited) that represents storage dams with the image of a faucet! (Alcan, Montreal, ALBU-60-02)

Site of the Lac Manouan storage dam prior to construction. The Lac Manouan storage dam was built of rock and timber during the winter months of 1940–41 in order to meet British contracts for aluminum ingot. The 1,250-foot-long dam would be built at the rapid in the foreground in order to enlarge and impound the expansive chain of lakes seen beyond. (Rio Tinto Alcan, Montreal, 00076 10)

Lac Manouan construction by airlift. Lacking the time to build an access road to the remote construction site, the Aluminum Company of Canada airlifted all necessary workers and equipment. (Rio Tinto Alcan, Montreal, CDP-40-1 #2)

Burning operations, August 1940. Fire was used to clear the dam site. A portion of the workers' camp can also be seen. (Rio Tinto Alcan, Montreal, CDP-40-1 #17)

Timber cribs under construction, October 1940. Workers build the timber cribs that form the foundation of the dam. (Rio Tinto Alcan, Montreal, CDP-40-1 #10)

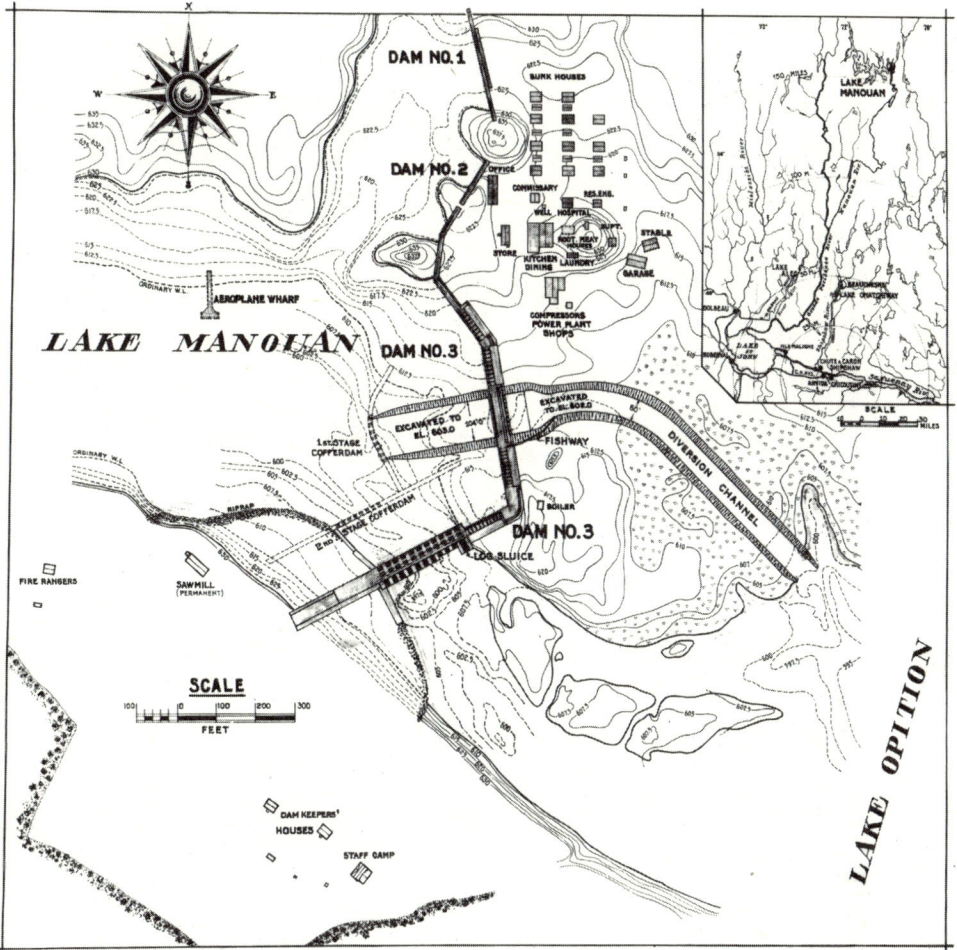

Lac Manouan project layout. In addition to showing the location of the dam, the schematic includes that of the construction diversion channel and workers' camp. (Rio Tinto Alcan, Montreal, 00076 10)

Lac Manouan dam completed. By the summer of 1941, the dam was completed and Aluminum Company of Canada controlled the flow of the Peribonka's major tributary. This photo looks upstream. (Rio Tinto Alcan, Montreal, ALBU 51-04)

Passe Dangereuse access road. The Passe Dangereuse reinforced-concrete storage dam, built in 1941–43 to meet US contracts for aluminum ingot, created what was then the largest reservoir by volume in the province of Quebec. Dam construction required a lengthy access road, seen here in winter use, which would open the Peribonka watershed to industrial civilization in the post-war years. (Rio Tinto Alcan, Montreal, CDP-43-02)

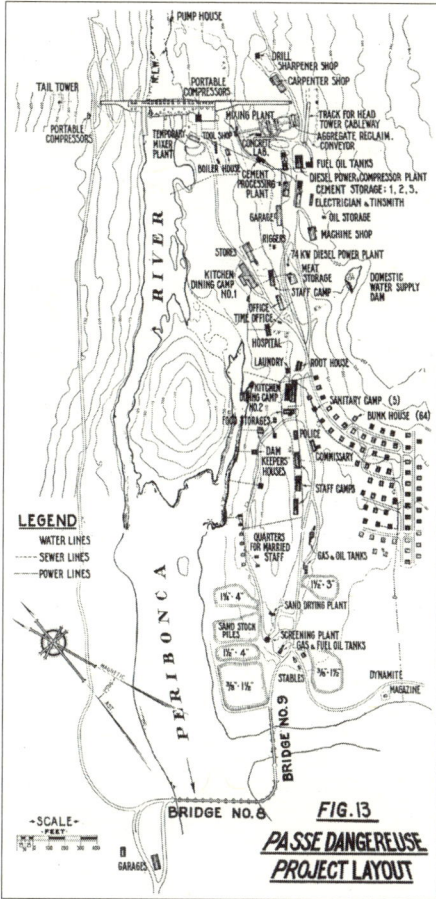

Passe Dangereuse project layout. The diagram shows the placement of the dam along with an extensive construction camp. At its peak, the workforce numbered 1,500 men. (Rio Tinto Alcan, Montreal, 00087 07)

Passe Dangereuse under construction. The nearly completed Passe Dangereuse dam, at over one hundred feet in height, would raise the Saguenay River's constant year-round flow volume nearly 10,000 cubic feet per second. The truck and workers' buildings at the base of the image provide a sense of scale. (B.J. McGuire and H.E. Freeman, "How the Saguenay River Serves Canada," *Canadian Geographical Journal* 35, 5 (1947): 200–25; photograph, 215. Photographer is unknown.)

Completed Passe Dangereuse dam, 1943. The reservoir is filling behind the completed dam in this aerial photograph of summer 1943. Swaths of uncut timber are partially submerged in the expanding lake. A company historian celebrated the project's achievement: "a little under twenty-one months after the first tractor attacked the access path, the Grand Peribonca River, the storied major tributary of Lake St. John, was bridled at Passes Dangereuses." (Rio Tinto Alcan, Montreal, CDP-43-13)

Aluminum ingot at Arvida Works. Additional storage provided additional power. Aluminum Company of Canada would supply roughly one-third of Allied aluminum ingot during the course of the Second World War. Here, cast aluminum ingots at the Arvida Works along the lower Saguenay River sit ready for shipment. (B.J. McGuire and H.E. Freeman, "How the Saguenay River Serves Canada," *Canadian Geographical Journal* 35, 5 (1947): 200–25; photograph, 216. Photographer is unknown.)

3

Passe Dangereuse

While Lac Manouan was constructed to serve British demand for ingot, the reservoir on the upper Peribonka River at Passe Dangereuse – the largest by volume in the province to this date – was created to serve growing wartime energy and ingot demand in the United States. The story of US power demand for aluminum manufacture and the inception of the Shipshaw hydroelectric plant is more fully documented in the next chapter.[1] In brief: it was late in the autumn of 1940, with the Lac Manouan dam under construction, that Canada's power controller, Herbert Symington, began urging Aluminum Company to complete the hydroelectric development of the Saguenay River as a means of meeting North American energy demand. Shortly thereafter, the US government sought additional supplies of aluminum ingot from Aluminum Company of Canada; the first contract between the company and the US government's Metals Reserve Company (a subsidiary of Reconstruction Finance Corporation, a federal government agency) was dated 2 May 1941. Alcan historian Paul Clark linked this to the Saguenay and Peribonka projects thus: "when Washington requisitioned its tremendous tonnages, more storage for more generators became urgent."[2] The result was the gigantic Shipshaw project on the lower Saguenay River, as well as the need for additional measures to steady the Saguenay River's volume of flow.

We should note that this cause-and-effect relationship was not as tidy as it appears in retrospect. Even before Shipshaw was definitively decided, Aluminum had pursued the possibility of developing the upper Peribonka – whether for more water storage or for electrical power, or both. The company decided to proceed with Shipshaw in April or early May 1941, while talks with Quebec over the Peribonka began as early as January. So the negotiations to dam at Passe Dangereuse predated Shipshaw by several months – during which time Aluminum Company

contemplated other power schemes, short of Shipshaw, before going forward with the enormous works on the lower Saguenay. Passe Dangereuse
was among these alternative and relatively small-scale projects. Company
engineers had already identified the site in the fall of 1937 on their
return from the surveys of Lac Manouan: a fifteen-mile-long gorge,
located thirty-five miles upstream of the confluence of the Manouan
River, through which the Peribonka River tumbled 600 feet. The length
and ferocity of the rapids in this stretch had long forced native trappers
to use an alternate route to ascend and descend the river in this vicinity,
and had also given the site its name. Aluminum Company engineers had
begun measuring the river's flow in the fall of 1940 to determine how
much water could be impounded and how much power damming would
yield. By the winter of 1940–41, they were planning the design of the
dam and surveying the contours of the projected flood line behind it.[3]

The record shows that the politics of Passe Dangereuse began in
the opening weeks of 1941 when Powell met with Premier Godbout to
plead his company's case for further expansion in the Saguenay region.
Aluminum's president made clear that his company had established
itself in Quebec primarily to exploit the province's abundant power
resources. He touted the company's contributions to both the economic life of the province and the war. Using the argument employed
by engineers and conservationists since the turn of the century, Powell
told the premier, "water power is no good if not used." And he insisted
that in order to compete in the post-war period, the company wanted
additional waterpowers "under its proper control" rather than simply
leased from the province. Acknowledging that the storage reservoir at
the Peribonka would flood considerable tracts of merchantable timber,
perhaps a million cords, Powell nevertheless pressed the premier for
his co-operation in this and other tasks that required the authorization
of the state. Powell bluntly said: "We want action."[4] Godbout was apparently willing to co-operate. But not to be rushed, the premier referred
Aluminum Company to the "proper parties" of the Department of Lands
and Forests.[5] Latreille and Dubose were in conversation by the end of
January, as Saguenay Power's superintendent assembled the data to
make an application. Latreille, for his part, reminded Dubose of the
need to obtain prior approval from the provincial Legislature if it was
a power project that the company intended, a requirement of Quebec
law since 1935.[6]

What we might describe as preliminary negotiations for Passe
Dangereuse ran from January through early April, in the period when

Aluminum had not yet firmly decided to build the Shipshaw dam. Aluminum's attorney J.A. Prud'homme prepared a draft bill for submission to the Legislature, which in turn was read and adjusted by the Hydraulic Service chief Latreille.[7] By early March, Dubose submitted a formal request for a concession: permission "to purchase" a portion of the Peribonka's bed and powers at Passe Dangereuse where the company would construct a dam and power plant utilizing 140 feet of head and generating 75,000 horsepower. The company planned to complete the storage dam in time to capture the spring runoff of 1943, with the power plant generating electricity at its maximum capacity by 1946. The electricity would be conveyed to Isle Maligne by transmission line, and the extensive 'pond' would be used to further regulate the flow of the entire watershed. Unlike at Lac Manouan or Lac St-Jean, the reservoir did not use a pre-existing lake or chain of lakes, so considerable land would be submerged. Of some 110,000 acres, the majority was Crown land, with the remainder held by a paper company. Expected to contain 225 billion cubic feet of water, such a reservoir would be by far the largest by volume in Quebec.[8] The next largest body, the vast Gouin reservoir in the St Maurice basin, came in at 180 billion cubic feet; Lac Manouan, broad but relatively shallow, was one-fourth the volume of its Peribonka neighbor.[9]

To the Hydraulic Service, this enormous body of water promised rich revenues for the province. In multiple memoranda, the Service's engineers dreamt of dollars: $70,000 annually in power taxes at Isle Maligne (due to (a) the expected 80,000 horsepower increase from the increased flow, taxed at the 50 cents per horsepower, and (b) the supplementary royalty on Lac St-Jean storage) and roughly this same amount ($72,000) at Chute-à-Caron, where industry had never before paid a power tax on the stored water in Lac St-Jean; a couple of thousand dollars in rent of Crown land (using the province's standard formula of 25 cents per acre for the lands to support the dam and accompanying structures); and a yet-to-be-determined dollar amount in compensation for the destroyed timber, which Hydraulic Service chief Latreille estimated at $65,000. And further revenue would be forthcoming in the form of power taxes if Aluminum Company chose to construct the new power plant at Passe Dangereuse in addition to the reservoir: at the current rate of $1.00 per generated horsepower, the 75,000-horsepower plant could yield $75,000 annually for Quebec.[10] Latreille was firmly opposed to an outright sale of the Peribonka's bed and waterpowers, as no sale of waterpowers had been accorded private industry for twenty-five years.

Although Aluminum's predecessors had purchased the Saguenay's powers in this fashion at the turn of the century,[11] these historical events in no way guaranteed the company vested rights to the remainder of the basin. Latreille was also determined that the state reserve the right to control the flow of the Peribonka (as in the case of the Manouan) in order to serve any future hydroelectric plants along the lower falls of the river.[12]

Minister of lands and forests Côté conveyed these recommendations to both Premier Godbout[13] and the officials of Aluminum Company. Côté met with Aluminum's attorney/corporate director Aimé Geoffrion in the office of the premier in early April and also separately with Saguenay Power superintendent Dubose. The minister offered the company a long-term, emphyteutic lease of forty years, annual rent of Crown lands at $2,000, and power taxes at 50 cents per horsepower at both Isle Maligne and Chute-à-Caron sites (not including the supplementary taxes to be paid on the 1922 agreement regarding Lac St-Jean). In the case of a waterpower development only, with limited storage, Quebec asked Aluminum for $1.00 per horsepower generated per year; and in the case of combination of power and storage, the above-mentioned rates in the proportion applicable. The state reserved the right to regulate the flow of the Peribonka "if and when deemed necessary," and noted that none of these offered terms could be deemed final until Aluminum Company proposed its construction program to the department "in definite form." Aluminum would also have to reach satisfactory agreements with the companies owning the timber. "I am confident," Côté closed his letter to Dubose, "that you will realize the necessity for us of safeguarding both public and private interests, and ... considering the extent of lands and rights affected as well as the increase of revenues made available to your Company ... you will find the above conditions to be most reasonable and acceptable."[14]

Aluminum's executives did not. Powell sought and gained another audience with the provincial Cabinet where Aluminum's president pressed his plan to purchase and dam Passe Dangereuse.[15] As he was unsuccessful in swaying the premier to his demands, Dubose threatened to drop the power project on the Peribonka "if it must be subject to an Emphyteutic Lease." In the matter of a storage development, he insisted on a seventy-five-year period (rather than forty) as well as an upper limit on royalty payments of 75 cents per horsepower/year made available at the Saguenay power plants. Latreille was reticent but ultimately willing to grant a seventy-five-year lease, although forty years was closer to standard practice, and the majority of provincial storage reservoirs (as in the case of the Gouin reservoir) were owned and controlled by the government. Côté and Latreille could agree to an upper limit on the power tax. In both of these instances, Côté convened a special meeting of the provincial Cabinet to approve the adjusted terms. But neither the chief engineer nor the minister would consider selling, rather than leasing, the Peribonka's riverbed and waterpowers.[16]

Aluminum, in any case, now had other reasons to give up the Passe Dangereuse power project. Even as Dubose was negotiating in Quebec City, his superior Powell was in Washington, DC, along with chief salesman MacDowell, negotiating a first ingot contract with officials of the United States government's Metals Reserve Company, one of the eight wartime subsidiaries of the government's Reconstruction Finance Corporation engaged in war production. This contract stipulated a manner of funding construction of the giant Shipshaw dam on the lower Saguenay River with the capacity to generate half a million horsepower and, shortly thereafter, a full million: the US government would advance $68.5 million against the future delivery of ingot.[17] For Aluminum Company, storing and gradually releasing water in the Peribonka basin at Passe Dangereuse would be an essential feature of this enormous hydroelectric project, whereas building the Peribonka power plant would not. By late April, on the verge of completing the US government contract, Aluminum Company abandoned its intention to build a power plant on the upper Peribonka. No advance authorization by the Legislature would be required. The company intended to build a storage reservoir exclusively at Passe Dangereuse. Impounded by a dam a little over one hundred feet in height, the reservoir would snake upstream some sixty miles between the canyon-like hills of the upper Peribonka valley and raise the Saguenay River's year-round flow volume nearly 10,000 cubic feet per second.[18]

With Aluminum's plans now defined, minister Côté felt competing pressures in the matter of a concession. In early May, Powell phoned the minister from Washington, informing him of the enormous, and now completed, US contract for ingot as well as Aluminum's go-ahead on Shipshaw, and pressing the company's now urgent need to create a storage reservoir on the Peribonka.[19]

While Powell attempted to expedite the political process, the principal players of Quebec's hydro bureaucracy sought to increase its rigour. Hydraulic Service chief Latreille, in particular, urged his superiors to exercise "great circumspection" in the matter of Passe Dangereuse. The

Lac Manouan accord, he noted, gave up significant revenues that might have accrued to the state. So did the 1922 agreement for the use of Lac St-Jean when it allowed the Duke-Price interests to pay taxes only on energy produced in excess of 200,000 horsepower.[20] Of greatest advantage to the company were the original turn-of-century sales of the bed and powers of the great Saguenay River, which Aluminum acquired from its predecessors, and from which the government earned no revenue whatsoever. From the 800,000 horsepower currently developed at Isle Maligne and Chute-à-Caron, the state might be garnering $400,000 each year. "In sum, according to the terms of existing contracts on the Saguenay River, the Aluminum Company of Canada and its subsidiaries has at its disposal 805,000 installed horsepower, having produced some 600,000 horsepower in 1940. It benefits from storage in the greatest reservoir of this Province and pays for these privileges but an annual sum of some $88,000, being less that 11 cents per horsepower installed or some 15 cents per horsepower produced." And now the company proposed to build the enormous power dam, Shipshaw, and to construct the largest reservoir in the province on the Peribonka River. Certainly the firm had spent considerable sums developing industry in the Saguenay region, Latreille acknowledged, and, yes, aluminum was an essential strategic material in the current war, but the company's profits and privileges had also been considerable, Latreille concluded. The "favourable treatment by which the company has benefited in the past must induce us, from this point forward, to be extremely cautious in granting other reductions or exemptions of payment."[21]

The head of the Quebec Streams Commission, engineer Olivier Lefebvre, also counselled caution, but he did so in accordance with the mission of his organization. While the Hydraulic Service was formed in 1910 to better regulate and manage the exploitation of hydraulic resources by the private sector, Lefebvre's organization, formed the same year, served as the province's technical advisory board in the field of hydraulic engineering, also taking responsibility for the construction of government-operated storage reservoirs. In short, the Streams Commission, like a Tennessee Valley Authority in miniature, represented Quebec's relatively minor concession to public ownership during the Progressive Era. Lefebvre had now caught wind in the local press of what was to be the province's largest reservoir. Not surprisingly, he sought a role for the Commission in its management. Lefebvre suggested that the Streams Commission approve the technical drawings and plans for the dam, and also oversee Aluminum's construction program. In addition, he urged that the province follow the examples of reservoirs that had been constructed at private expense on the Gatineau, Lièvre, and Mattawin rivers, but then reverted to being the property of the government and operated by the Streams Commission. At Passe Dangereuse, this would mean that in lieu of the government leasing the lands and storage rights to Aluminum Company, Passe Dangereuse would be built by Aluminum but retained and managed by the state. Here was an interesting divergence of opinion between the Service and the Commission, reflecting different possibilities for government intervention in what was still a privately dominated hydroelectric industry. Quite soon, in 1944, Lefebvre's mission to expand public ownership would make an important political stride in the formation of Hydro-Québec. At the Saguenay, however, private enterprise would prevail. Côté was content to bring Lefebvre up to date on the early stages of negotiation and to involve the Commission in the approval of plans and the oversight of construction. On the larger question of state control, the minister did not respond.[22]

Others in the province – third parties affected by the project – also reacted to the news of the impending reservoir. Farmers, we should note, were not among these: there was no agricultural land at Passe Dangereuse. Latreille well understood that this was a region "far from the centres of colonization," where the interests of rural people would not require mediation.[23] In this sense, Passe Dangereuse, despite its enormous scale, would be immune from the intense and primal conflict that had accompanied the impoundment of Lac St-Jean. The conflict was certainly still fresh in memory for those who had experienced it.[24] Moreover, both the state and company were aware that the reservoirs might actually offer some conciliation to the populace downstream, since regulating the lake's major tributaries would tend to 'flatten' the flood runoff, in turn reducing seasonal flooding of low-lying farmland around Lac St-Jean. Far from raising the ire of farmers, building reservoirs in the Peribonka's uplands might help to abate their discontent. Latreille himself, as we have seen, had urged in the late 1920s that the state form reservoirs in the Peribonka basin as a means of flood control around Lac St-Jean.

There were of course other resources and other interests on the Canadian Shield. A gold prospector from Chicoutimi wrote to Lands and Forests to express concern that the flooded land would prevent his mining operations in the region. This single letter triggered a small flurry of activity inside the Quebec government, at the centre of which was Latreille, as befit the Hydraulic Service's function as the intermediary between private and public interests. The department soon informed the prospector that Aluminum Company had not yet obtained the government's approval. But if or when authorization was given, "you can be assured that equitable compensation would be paid by the company to all third parties that would be affected by these works." To prevent any further possible conflict between the big hydroelectric project and the interests of miners, and to protect Aluminum from legal claims in this resource domain, the Ministry of Mines crafted and passed an Order-in-Council closing the reservoir site to further prospecting and staking.[25]

Potential conflict over timber would not be settled so easily. Aluminum Company pursued its surveys of the reservoir's flood line through the winter and spring of 1941. By summer, with the high-water contour almost entirely drawn on the map, it was plain that the Passe Dangereuse dam, at its ultimate height,[26] would submerge a very considerable expanse of merchantable wood: roughly thirty square miles of Crown timber, forty-five square miles of Crown timber limits leased by the Consolidated Paper Corporation of Montreal (formerly Port Alfred Pulp and Paper), and forty-eight square miles of freehold timber lands owned by the Montreal-based Quebec Pulp and Paper Corporation (formerly Compagnie de Pulpe de Chicoutimi). Both paper companies were controlled by English-speaking Canadian investors and had been significant corporate players in the Saguenay region's pulp and paper industry.[27] Lands and Forests estimated that, in all, the dam would flood 200,000 cords of government wood, as well as at least 128,000 cords on the privately held land of Quebec Pulp and Paper. Neither of the private firms had yet exploited these most northerly of their woodlands; nor, by their own estimates, would they exhaust their more-southerly stands of timber for fifteen, or even thirty, years. Still, Quebec Pulp lands had been at least superficially surveyed in the 1929–30 season, with forestry engineer G.H. Gustafson reporting mature spruce "in good condition ... over the whole territory," which comprised 4.1 million cords of wood.[28] Both Quebec Pulp and Consolidated Paper held valid legal claim to the use of the land and some compensation was due. Thus the Hydraulic Service's Latreille took care to meet and correspond with the executives of the respective firms to inform them of the hydro project, to get them the data required to negotiate a compensatory arrangement with Aluminum, and to share the relevant clauses of the Order-in-Council as they were drafted and redrafted. The stumpage prices were settled quickly enough: $1.00 per cord for the state's timber (reduced from

$1.25 per cord in consideration that Aluminum would turn over its new fifty-eight-mile access road to the government), and $2.50 stumpage per cord for the lost wood of Quebec Pulp and Paper, provided that Aluminum also pay the costs of harvesting and removing the timber prior to constructing the dam.[29]

These negotiations soon hit several snags. First, early in 1942, with dam construction underway, Aluminum Company informed the government that it would not be able to salvage all the trees from the reservoir site. Aluminum's attorneys claimed a practical shortage of available manpower during the war and, more significantly, a dire need to complete the main dam in time to capture the spring freshet of 1943 in order to meet Allied demands for ingot and airplanes. In this sense, argued attorney J. Alexandre Prud'homme, the loss of some 120,000 cords of wood would, however, mean manufacture of 1,700 combat aircraft and "a precious gain" for the war effort.[30] To the timber firms, leaving a partly submerged forest meant obstructions to future log-driving operations, at least until the annual action of ice eventually cleared the reservoir. For the Department of Lands and Forests, there were also political ramifications in permitting Aluminum to leave this wood to rot beneath the waters of the artificial lake. As under-minister, Bédard asked his superiors: "How will public opinion react when it's learned that we've let a considerable volume of wood perish that might otherwise have been exploited?"[31] Minister Côté agreed that the matter was "very important," and thus brought it to the attention of Godbout and the Cabinet for a decision.[32] In the wake of this meeting, the state renegotiated additional financial compensation to let dam construction go forward – Aluminum would pay the state an additional 6.25 cents per cord on the uncut trees, or some $60,000. And by mutual agreement, a committee of three representatives (of the ministry, Consolidated Paper, and Aluminum) would determine "when and to what extent it may be necessary to clear the river banks of the tributaries [of the Peribonka, and] to clear a logging channel" through the reservoir.[33]

The second dispute also concerned the future passage of logs in the Peribonka valley. Both Quebec Pulp and Paper and Consolidated Paper insisted that Aluminum pay not only for the lost wood but also for the costs of necessary upstream improvements – including a well-placed sluice gate in the dam to accommodate logs, and multiple piers and booms in the reservoir above in order to conduct and corral them – to ensure the wood's water-borne transportation. Aluminum balked at these additional investments. Company secretary V. Edward Bird expressed "astonishment" at the timber companies' demands since "neither of the companies ... has ever conducted a pulpwood operation in the area in question," and countered that, "Our construction programme has greatly facilitated the conduct of logging operations in the area ... by providing an access road and a dam to control the flow of water." [34] It would be just as fitting, wrote Dubose, that Aluminum "collect a toll" from the timber interests for use of this access road as for the dam-builders to shoulder the cost of upstream improvements for the passage of wood. It was the access road, after all, Dubose wrote presciently, that "will open up a timber area below Passe Dangereuse which would supply the needs of all the mills in the Saguenay for many years."[35] No agreement could be reached between Aluminum and timber companies over these arrangements prior to the passage of the Passe Dangereuse concession. As in the case of wood salvaging in the flood zone, it was left for the representative committee established by the eventual Order-in-Council to "determine when and what improvements, if any, shall be made in the flooded area ... also when and what equipment ... to establish that logging operations may in the future be conducted as effectively as would have been possible under normal conditions prior to the construction of said dam."[36] Such conditional language, suggested by Aluminum's executives, could only favour Aluminum's interests in future negotiations.

In the course of two years' squabbling over timber, these two major points of disagreement were accompanied by several smaller ones regarding the quantity as well as the quality of the wood for which Aluminum should pay dues, as well as whether Aluminum had been fully forthcoming with the timber companies as to the eventual size and extent of the reservoir. Throughout these talks, several patterns emerge. First, there was never a moment when Big Timber might have indefinitely blocked or prevented the construction of the Passe Dangereuse dam: in this sense, the war emergency trumped timber in the Passe Dangereuse talks. As Quebec's under-minister for lands and forests put it to Aluminum Company's corporate secretary in not-quite-fluent English: "we don't want to put any obstacle to your project."[37] At the same time, it was always certain that financial compensation was forthcoming, that money would change hands. The amounts were ultimately determined by government's review, totalling $320,000 for Crown land timber and $504,000 for that of Quebec Pulp and Paper: in this manner forestry interests were adequately served and well represented in this political process.[38] Finally, we should acknowledge the role played by the state's

hydro bureaucracy. Latreille's Hydraulic Service performed its function effectively and with a measure of impartiality as the official conduit
and intermediary between the hydroelectric developer and third-party
industrial interests: informing, conveying, conjoining, and cajoling in its
efforts to urge the disparate interests in this common northern resource
to reach a mutually satisfactory bargain. The timber companies themselves, in the heat of these negotiations, characterized the state's role as
"useful intervention."[39]

Latreille's Hydraulic Service performed a similar function as intermediary in the matter of the dam's effect on fish. By very recent Quebec
law, any obstacle to be placed in a waterway must be approved by the provincial Ministry of Fish and Game (Chasse et Pêche), the minister either
attesting to his satisfaction that a migratory fish passage was provided
or recommending that such a fishway was not appropriate.[40] Fishways
being something of a novelty in the world of hydraulic engineering, first
developed and tested on western salmon rivers,[41] Quebec, like other
regions of North America, was still experimenting with their use and
regulation. Latreille conveyed plans for the Passe Dangereuse dam to
the Ministry of Fish and Game in September 1941 requesting advice in
this regard;[42] from this point forward, the under-minister of fish and
game, L.A. Richard, made life difficult for the officials in Aluminum
Company who sought a provincial concession.

As the dam would certainly affect game fishing, Richard argued, it
would affect the "tourists" that visited the region.[43] When Aluminum
countered that tourists did not get anywhere near Passe Dangereuse
and that the river in this section "has never been traversed by human
beings,"[44] Richard changed his tack. Perhaps, he acknowledged, no tourists visited the upper Peribonka valley, but "all dams affect fisheries,"
thus Aluminum was obliged to construct a migratory fish ladder, one
that had been designed by expert biologists.[45]

Richard engaged the well-known, if eccentric, architect and sport
fishing advocate Percy Erskine Nobbs to propose a program of mitigation and to design the fishway. Scottish-born Nobbs, professor of
architecture at McGill University, passionate fisherman, and author of
a fishing manual, was hardly a neutral observer. He opposed "the damming and pollution of rivers" as a clear threat to sport fishermen and he
deemed Quebec a particularly poor example of fish conservation. "The
salmon, like the white pine, has been nearly squandered out of existence
in the Province of Quebec," he had written in his 1935 book *Salmon
Tactics,* and "even since the war several good rivers have been ruined by

dams for pulp and power operations without the least regard to the question of fishpasses."[46] For the study of the Peribonka, Richard's ministry had selected Nobbs to carry out the work, while his expenses were being charged to Aluminum Company, as would be the costs of mitigation. In February 1942, Nobbs penned a preliminary report, "Anticipated Effects of the Passe Dangereuse Dam." As the report was based on but a "second hand" study of available maps and plans showing the watershed and the Passe Dangereuse reservoir, Nobbs could only vaguely predict "prejudicial effect on the fisheries of the whole of the main river with diminution in the lower reaches"; and he called for an on-site survey of the basin. By October, with the aid of recognized biologist and ichthyologist Ukrainian-born Vadim D. Vladykov and a survey team, Nobbs had made a visit to and a survey of the Manouan River, which was already dammed. He could, therefore, report more definitively on the probable effects of blocking the Peribonka's main trunk.

Nobbs's final report was darkly pessimistic. "When I made my [first] report," he noted, "I was in ignorance of the existence of the Manouan Dam which went into operation in April 1941." Its consequences included masses of "decayed vegetable matter" derived from the reservoir's submerged banks, which "blackens one's teeth in a few days," while the "loss of the spring flood ... upsets a condition ... which has existed for millions of years." The combination of such "pollution" from reservoir flooding and the changed flow rhythm of the river, in addition to changes in water temperature and winter ice conditions, had produced "a great reduction of stock" in the Manouan valley, Nobbs asserted. This was especially true for the landlocked salmon species ouananiche, which spawned in the Peribonka's tributaries and descended to Lac St-Jean to feed. Given the larger scale of the Peribonka project, which he understood to be a one-hundred-foot-high dam and sixty-mile-long reservoir, the effects of Passe Dangereuse would undoubtedly be "far more pronounced," Nobbs predicted. "The important ouananiche fishery which is based on the Peribonka and Manouan spawning grounds appears to be doomed," he wrote, and no amount of money could compensate for these losses, nor would a fish ladder or fish hatcheries "be of the slightest use."[47]

We should note that Aluminum Company had begun to encounter, and to try to counteract, charges by local residents that the company's Isle Maligne dam at the outlet of Lac St-Jean was damaging the renowned ouananiche sport fishery and thus the region's tourist trade.[48] Aluminum's response to Nobbs's findings regarding the Peribonka was practical and callous: since a fish ladder wouldn't help the fish, they suggested simply jettisoning this particular clause in the Order-in-Council and getting on with the project.[49]

By early 1943, when all other issues of the concession had been resolved, the issue of fisheries remained "the sole point to debate."[50] A change in personnel – appointment of a new minister of fish and game in the Godbout administration – had delayed resolution of this problem. Moreover, Aluminum's intention to leave the majority of the timber standing led Fish and Game experts to assert further future damage to fisheries. Richard continued to insist on a fish ladder designed by experts, even though he acknowledged that the fishway would not likely be an effective remedy. Richard also wanted extensive future studies of the dam's effects on fish that would be funded by Aluminum Company, as Percy Nobbs had recommended, and wanted to ensure, by the language of the Order-in-Council, that the company would co-operate in this regard. Aluminum continued to balk at what its executives considered unnecessary expenses and arrangements. Latreille grew somewhat frustrated with the disparate attitudes of the two parties, both with Aluminum's "over aggressiveness" on the one hand and Richard's "righteous indignation" on the other.[51] These events led the Hydraulic Service chief to suggest changes in Quebec law to the new minister of lands and forests, Wilfrid Hamel, that would bring Quebec in line with Ottawa on the issue. Latreille's intention was to acknowledge and deal with the inevitable damage to fisheries from a hydroelectric project – if a fishway was not required, then financial compensation from the dam builder that could be used to support hatcheries – but without giving the Ministry of Fish and Game the ability to indefinitely delay authorization for such an important industrial project.[52] In the end, the Order-in-Council was written in this spirit. No fishway was required of Aluminum Company, as "the Minister of Fish and Game has expressed an opinion to the effect that a migratory fishway would not be the proper remedy in this particular case." Yet the state retained the right to "impose on the petitioner an equitable compensation for the prejudicial effects caused to fisheries," which was then fixed at $50,000.[53] This marked the first instance in the province of Quebec of compensation being paid for damage to fisheries.[54]

It should be noted that fisheries, rather than fish, were of concern to Quebec's functionaries and biologists. The human use of this resource, rather than ecological damage per se, constituted the primary consideration in this pre-dawn period of the modern environmental era[55] – and the particular humans that L.A. Richard and Percy Nobbs had in mind were well-to-do sportsmen from cities like New York and Montreal. That human use was also relatively minor, in the sense that mid-twentieth century tourism in the Lac St-Jean region didn't bulk nearly as large in economic importance as electrochemical manufacture (let alone pulp and paper production).[56] In short, the Saguenay did not support a major salmon fishery as did rivers such as the Fraser or Columbia. The Godbout administration saw Passe Dangereuse, regardless of its economic contribution, as a component of Quebec's war effort. Concerns for fisheries, therefore, would not be allowed to interfere with the construction of the dam. During wartime, in this Eastern Canadian contest of "fish versus power,"[57] hydroelectric energy was to be the clear and predetermined winner. As Latreille explained appeasingly to Richard: the "necessities of the hour oblige us to consider favourably certain projects that, at other times, would have been rejected upon presentation."[58] Still, like forestry, fisheries were involved in, as well as financially compensated by, the political process. And in this political dialogue, it was the Hydraulic Service that again acted as relatively neutral protector of third-party interests, albeit with a penchant to promote industrial and, particularly, hydroelectric development.

While a fisheries clause formed the last obstacle to a concession, it was hardly the most difficult or significant one for Aluminum Company and the Quebec state to overcome. Latreille and Dubose had roughly come to terms with the amount of the power tax during the initial stage of Passe Dangereuse's negotiations. But as Aluminum made ready to submit a formal application for a concession in the summer of 1941, still to be determined was the seasonal period on which that tax would apply, and how the tax was to be calculated. Both factors were financially significant to the ultimate charges to be paid by Aluminum to the province during the life of the lease.[59]

The power tax was the major bone of contention in the Passe Dangereuse talks. At first, Latreille insisted that this royalty should be a fixed charge of 50 cents per horsepower "based on the maximum power which might be generated with the stored water." This was the typical arrangement in Quebec, relying on standard and easily calculated ratios of flow volume to horsepower production; it also promised maximum revenues to the state. Aluminum countered that fixed charges would excessively burden the company in the uncertain post-war period and that the manufacturer should not be held to precedents that applied largely to "public utilities" that generated electricity for consumer use.

Aluminum appealed to Minister Côté, who refused to rule against his department's engineers but agreed to arrange a hearing in the Cabinet. In early September 1941, Aluminum's attorney J. Alexandre Prud'homme made the case before Quebec's cabinet as to why his client should pay royalties on power only when the Passe Dangereuse reservoir was in active use. The Cabinet, like Côté, was noncommittal, expressing only its desire that the parties "work out some plan which both the Aluminum Company and the Department would recommend." Latreille, seeking to honour the Cabinet's request but unwilling to lose a dime of revenue, suggested a compromise: in lieu of paying the standard tax of 50 cents per horsepower *stored*, Aluminum Company pay $1.00 per horsepower actually *generated* downstream at the several hydroelectric plants along the Saguenay River during the four-and-a-half month winter season. The resulting charges would be similar and no precedent would be set.[60]

Calculating this tax would hardly be a simple matter. Passe Dangereuse was being created to augment the flow of a watershed that was already being regulated by two other reservoirs, Lac St-Jean and Lac Manouan. How exactly would one distinguish between the energy-generating benefits of the new reservoir versus the existing ones? Or, as Latreille and Dubose saw the problem, above what "standard quantity" of horsepower already available in the Saguenay's hydraulic system was it legitimate to charge Aluminum Company for the use of a provincial resource?

This debate over the "standard quantity" occupied Latreille and Dubose in long and rigorous negotiations through the autumn months of 1941 and beyond. Both engineers pushed the best possible financial arrangement for their respective interests. And whatever their private disgruntlements with the talks, each publicly maintained a tone of considerable mutual respect. "Dear Mr Latreille," wrote the gentile and articulate Carolinian early in 1942 following a holiday break in the talks. "I hope you had a nice Christmas and New Year, and are ready to face 1942 with plenty of vim and vigor – not that I want you to exert it all on me however." "Dear Mr Dubose," replied Latreille in his perfectly fluent English, "I enjoyed the usual Christmas and New Year's let up and I presume that for the benefit of your family and friends, you succeeded in doing the same. Hence, you had to divert momentarily your dynamic energy to more enjoyable occupations than the pursuit of your Company's plans. Let us hope that 1942 will not find us too often in opposition and in need of a referee."[61]

By this point, a referee was in fact required. Latreille calculated that Aluminum could generate 386,000 horsepower at its Saguenay plants during the winter months without the use of Passe Dangereuse. He was willing to revise this figure upwards in the lease, as high as 400,000 horsepower, if the company would consent to forego the ceiling of 75 cents per horsepower in royalties and/or adjust the royalty upwards after an initial ten-year rather than twenty-year period. Given his earlier objections to the "favourable treatment" experienced by Aluminum since the turn of the century, as well as his sense of the "privileges" already granted in the matter of Passe Dangereuse (a seventy-five-year lease and maximum royalty charges of 75 cents per horsepower), Latreille was unwilling to budge further. Dubose, for his part, would go no lower than 415,000 horsepower, claiming that he could not convince his principals to accept a reduction of this figure. This gap of 15,000 horsepower was not insignificant, as it represented $15,000 annually to be paid to the Quebec state. Latreille and Dubose had reached something of an impasse.

At Minister Côté's suggestion, Quebec Streams Commission's Lefebvre was brought into the discussions.[62] The Streams Commission was already involved in reviewing Aluminum's technical drawings and plans with respect to the stability of the dam, the capacity of the gates, and so on, and a Commission engineer was on-site at Passe Dangereuse to oversee construction. Now, Lefebvre would act as an arbiter, engaged by Côté to pore over the calculations of the "standard quantity" that had been made by the respective parties and to judge which method seemed the most accurate. As the months slipped by, however, Lefebvre's participation did not clarify the calculation. In the interim, Aluminum decided to increase the power-generating capacity of the Shipshaw power station from six to ten or even twelve turbine/generator units, which revised upwards the available power of the Saguenay system and thus the "standard quantity." And Hamel, long-time champion of public power, replaced P.-E. Côté as Godbout's minister of lands and forests.

Then in February 1943, with the Passe Dangereuse dam approaching completion, Aluminum's executives gained one more audience with the Cabinet to argue their case. Just before noon on February 2, president Powell, Dubose, Bird, and attorney Geoffrion presented the company's point of view before Godbout, Hamel, and the Cabinet on why Aluminum should not be taxed for power production when Passe Dangereuse's stored waters were not actually being released. In a separate session several hours later, Latreille, under-minister Bédard, and

Louis-Philippe Pigeon, procureur géneral (legal clerk) of the Quebec Legislature (and future Supreme Court member), reviewed the history of the negotiations to date and presented the government's case. Again, to the disappointment of Aluminum's executives, Godbout and the Cabinet did not overturn the decision of its bureaucrats. By late afternoon, Hamel was informing Aluminum that the government would ask $1.00 per horsepower generated at Isle Maligne, Chute-à–Caron, and Shipshaw by means of the new reservoir at a minimum amount of $25,000 per year "even if the reservoir wasn't utilized."[63]

By the end of February 1943, Latreille and Dubose and their respective colleagues finally came to terms over the complicated clauses regarding the power tax.[64] The government would grant a seventy-five-year lease "of all lands and rights of the Crown which are deemed necessary to the erection, maintenance and operation of the proposed storage dam and appurtenant works." In return, Aluminum would pay annual dues of $1.00 per horsepower per year on power generated in excess of a standard quantity of 408,000 horsepower (then 440,000 horsepower once nine or more units were installed in Shipshaw) generated collectively at Aluminum's power plants along the Saguenay River during the winter months. Minimum and maximum royalty payments ($25,000–$120,000) were also included, so that Aluminum would pay the province "whether the dam is used or not." Dues would be revised after the initial twenty years and then after subsequent ten-year periods; and the maximum per-horsepower tax would be $1.25 per horsepower, rather than the originally agreed 75 cents. If Aluminum released the waters of its reservoir prior to the usual low-water winter period, annual dues would apply regardless; i.e., Aluminum Company could not lower the reservoir in summer or autumn and then claim to pay no taxes for lack of use during the colder months.

In all, the government had held firm to its mission to produce a lease that would garner significant revenue, even while flexing its standard policy of taxing private reservoir use on a year-round basis. The accord was flawed in at least two ways, however. Latreille expected that the "long and tedious negotiations" regarding royalties would yield "an accord that would stand up to any circumstances."[65] In this, he was mistaken. The concession's "rather elaborate" clause dealing with royalties (as another Hydraulic Service engineer referred to it[66]) would later give the state considerable difficulty in calculating the annual dues owed by Aluminum. In June 1945, Aluminum submitted a cheque for $32,000 as dues for the 1944–45 period, calculated according to the company's own "interpretation" of the lease. The new chief of the Hydraulic Service, A. Euclide Paré, saw the calculation quite differently and called on Latreille and Yvon Deguise, now with Hydro-Québec, to advise. By late 1945, these two engineers were helping the Service to draw up a supplementary agreement to cover more explicitly the different possible uses of the reservoir and thus firmly fix the method of calculating the tax.[67]

More problematic in the short term was the fact that the Quebec government, under pressure of the war emergency, had allowed Aluminum Company to build an enormous dam on Crown land well before granting the formal authorization. The Passe Dangereuse concession, embodied in Order-in-Council 893, was passed by Quebec's Cabinet on 2 April 1943, and confirmed by Quebec's lieutenant-governor the following day.[68] Only then, by Quebec statute, could Aluminum commence to build its dam. The law in violation was chapter 46 of the Revised Statutes of Quebec, 1925 – referred to as the Water-Course Act – section 57 of which dealt with "The Construction and Maintenance of Reservoirs for the Storage of the Water in Lakes, Ponds, Rivers and Streams." The relevant section reads: "No work ... may be constructed or maintained, unless the plan and specifications relating thereto have previously been approved by the Lieutenant-Governor in Council." The penalty for dam-building without authorization, a harsh one to be sure, is also described here: "the demolition of such work ... may be ordered by any court of competent jurisdiction, upon an ordinary action instituted by the Crown or by any interested party."[69]

The historical record leaves no doubt that Aluminum dammed the Peribonka River before securing the legal authorization to do so. Aluminum's construction program had begun some twenty months prior to 1943's Order-in-Council 893 – in August 1941, when the company first awarded a contract to Dufresne Engineering Company to build the fifty-seven-and-a-half-mile access road through wild country from the northern terminus of an existing logging road. By the following December, the transportation contractor Price Brothers and Company began regular freighting by truck from the railhead at Dolbeau to the work site 135 miles north. For Alcan's historians this was an "epic story" in its own right, as several dozen trucks funnelled the myriad necessities of construction equipment, cement, dynamite, foodstuffs, and personnel. Timber cutting at the reservoir site was ongoing during the winter of 1942. So was preparatory construction for the storage dam itself: construction of rock-filled log-crib cofferdams, diversion sluices, a workers' camp of sixty-four log cabins, construction plant (complete with concrete processing and mixing plants, machine shop, compressor plant, carpenter shop, etc.) and a sizeable 1,600-foot-long cableway for the placement in the main dam of concrete, reinforcing steel, and various forms and gates. With preparatory work completed and the riverbed exposed by the summer of 1942, the erection of the storage dam began in earnest. At its peak, the workforce at Passe Dangereuse numbered 1,500 men. "In tribute to the local 'Blueberries'," described Alcan's Paul Clark with mild condescension, "they stuck to the job, behaved in decent fashion and seldom complained so long as, when the shift was finished, they were made comfortable." Aluminum's engineers closed the sluice gates of Passe Dangereuse on 4 May 1943 – almost exactly one month after the project was authorized – and the waters behind the dam began to rise. And so "a little under twenty-one months after the first tractor attacked the access path," Clark celebrated the company's impressive achievement, "the Grand Peribonca River, the storied major tributary of Lake St John, was bridled at Passes Dangereuses [sic]."[70]

It is less clear that this legal breach marked an extraordinary, or even unusual, historical event in early twentieth-century Quebec. As we have seen, with Lac Manouan, Alcan had duly awaited the government's Order-in-Council of August 1940 before launching construction of the storage dam. The other of the Saguenay basin's reservoirs, Lac St-Jean, possessed a more complex regulatory history: Alcan's predecessor company (J.B. Duke's Quebec Development Company) had also duly waited for government authorization in 1922 before damming the outflow of the lake; but, three years later, Premier Taschereau quite casually waived Duke's waiting period to raise Lac St-Jean to its full height, thereby permitting the Americans to flood the farmland ringing the shore and producing the "tragedie" of the 1920s.[71]

More generally informative in this regard may be a morsel of correspondence preserved in the Godbout Papers. Early in 1941, the formidable Montreal corporate attorney Aimé Geoffrion, an Alcan director, requested of Quebec's minister of lands and forests, Pierre Émile Côté, a change in Quebec law. Whether he acted on behalf of Alcan or another of his corporate clients isn't specified. Geoffrion sought an amendment to the Water-Course Act that would permit the government to approve hydroelectric works, whether powerhouses or reservoirs, after, as well as before, actual construction had begun. In asking to amend the law, Geoffrion assured the minister that he had already spoken with hydro bureaucrat Normandin of the Public Service Board

(and formerly of the Hydraulic Service), "who tells me that the practice
has *often* [italics added] been to approve after the fact; it seems that
the law should be amended to accord with its practice." Lest we doubt
the lawyer's truthfulness (since, of course, it behooved his argument
to cite a recognized government authority), we also have the Streams
Commission's Olivier Lefebvre writing in confirmation of Geoffrion's
position. Lefebvre to Côté: "To date, the department [of Lands and
Forests, via its Hydraulic Service] has approved these constructions after
the fact, but it doesn't seem that the law permits such approval. It is
desirable that the amendment suggested be inserted into the law."[72] In
fact, the amendment was not approved and did not enter the body of
Quebec law. Regardless, it is clear that Quebec's legal code was at odds
with actual practice regarding the concession of waters and waterpowers.
The Passe Dangereuse dam may hardly have been the only dam built in
the province prior to fulfilling the technical terms of the law.

Still, bending the law in this instance mattered. It certainly mattered
to Maurice Duplessis once he found out about it. Never one to let
factual accuracy or nuance get in the way of a political argument, the
Opposition leader would happily seize on the supposed illegality of the
Peribonka concessions to attack the government on its hydro policy,
as we will see. It mattered to Raymond Latreille, who required more
time to negotiate a concession. When Alcan began construction of Passe
Dangereuse in the summer of 1941, the Hydraulic Service chief still
needed many months, as it turned out, to work out terms regarding a
power tax as well as timber and fisheries.

And it mattered to Alcan. In order to store the water of the Peribonka,
to feed the giant turbines of Shipshaw by late 1943, Aluminum Company
required nearly two years lead time to erect the Passe Dangereuse dam.
Getting the construction program started early was essential; it also
involved risk. Alcan's attorneys obviously knew the law – as Geoffrion's
letter, cited above, makes clear; so did Alcan president Powell, as he
demonstrated in 1940 when he sought, unsuccessfully, to obtain from
Godbout "some form of authority" prior to an Order-in-Council to permit the corporation to commence dam construction at Lac Manouan.
No prudent corporation, whether in time of war or peace, would have
knowingly risked demolition or confiscation of its valuable property,
let alone a damaging blow to its already fragile political reputation
in the province as the power and aluminum 'trust.' In the minds of
many Quebecers, after all, 'Aluminum Company of Canada' was still
inseparable from 'Aluminum Company of America,' both of which were

closely associated with 'Duke-Price Power' and 'Quebec Development Company,' which had treated the local population so abusively in the 1920s. So it is very likely that Alcan asked for and received from Quebec's premier some form of guarantee of legal indemnity or release, some assurance of imminent government authorization, whether verbally or in writing, in lieu of an Order-in-Council. Where is the release in question? Godbout's papers in Quebec City don't yield such a document, nor do the records of the Hydraulic Service. The other obvious place to look would be in Montreal, in Alcan's board of directors' Minutes of the wartime years. Here and elsewhere in this study, the public sorely misses access to the corporate archives of Aluminum Company of Canada!

The available historical record also does not tell us just when Adélard Godbout quietly gave Aluminum the nod to build the dam. Unlike in the case of Lac Manouan, there is no written evidence of Aluminum approaching the premier for such unofficial authorization. It's possible that Geoffrion's request to amend the law was part of an effort to deal with this conundrum, but we can't be sure. Godbout's authorization to construct the Passe Dangereuse dam may have come early on, in the summer of 1941, when Aluminum first submitted its drawings and plans to the Hydraulic Service. As the company seemed "very anxious"[73] to get a concession at this time "on account of the emergency"[74] (to capture the spring runoff of 1943), we can imagine president Powell at this point begging the premier for his personal approval in order to get on with the work. Or perhaps the authorization occurred in stages, with Godbout permitting road construction only to go forward, in August, fully expecting that the unresolved terms over timber, fisheries, and royalties would be wrapped up in time for actual dam construction to commence by early 1942. Negotiations over each of these issues were prolonged, as we've seen. In the interim, Japan attacked Pearl Harbor, and the US entered the war. Perhaps it was only then that Godbout definitively cleared the way for a project that he deemed crucial to the war effort. This was, after all, now also the war effort of the United States, the country primarily for whom Aluminum was expanding its production of ingot in the Saguenay valley.

If Godbout himself had any doubts as to the necessity of rapidly damming the Peribonka to serve war production, he would certainly have heard from Ottawa. There is only suggestive evidence of federal intervention in the matter of the Peribonka River: rumours were aired by the Quebec press in the autumn of 1941 that Aluminum Company was attempting to mobilize federal ministers to press Godbout to make an "exception" in provincial hydro policy that would allow the company "to extend its monopoly at Lac St-Jean to all the rivers that feed this immense basin."[75] There is far more concrete evidence that federal officials conveyed their support for the hydroelectric expansion of the entire Saguenay watershed, which included the Peribonka. The Saguenay's exploitation was, after all, integral to a grand Canadian-American wartime bargain regarding electrical power flows and the production of aluminum ingot.

4

Beyond Quebec:
The Shipshaw-Massena Bargain

Little is written on the US electrical power industry during the Second World War. At war's end, the US government commissioned a broad array of historical reports and special studies dealing with the war-production effort, the purpose being a "full and objective analysis of the administrative problems and techniques involved in the mobilization of American industry" during the crisis.[1] Subjects ranged from truck manufacture to the labour shortage to War Production Board's policies toward lead and zinc. Electrical power was not among the topics, nor have scholars since made the effort to synthesize the war's electrical power history.[2] This is unfortunate, as power was not only critical to specific electrometallurgical industries such as aluminum and magnesium (which were themselves the objects of post-war study) but also served as the production matrix for many other industrial processes. "Modern warfare is a merciless competition between the industrial organizations and resources of the warring countries," explained the chief power consultant and later director of the war-production board, Julius Krug, early in the war. "The production of machines, measured in ships, tanks, planes, and guns, is the decisive factor in the destiny of nations. Aluminum, copper, steel, and chemicals are ordinarily considered the raw materials of war, and of defense. Electricity, however, is the common requirement of all." The *Wall Street Journal* echoed the sentiment: "Electricity is the keystone of the national defense program. Unless it is available in adequate quantity where and when wanted, fulfillment of the vast rearmament program would be inconceivable."[3]

Well before the United States entered the Second World War, the US government had begun planning to assure an adequate power supply for the nation in the event of a war emergency. In March 1938, President Roosevelt ordered the Federal Power Commission (FPC) to co-operate with the War Department in an inventory of the nation's power supply. The resulting report, submitted confidentially that summer, suggested "widespread and critical shortages" in the event of war, a situation that required "immediate attention." However, immediate attention was hardly the result. Through the spring of 1940, a series of government committees charged by Roosevelt with drafting legislation, including budget appropriations for new power plants, was hampered by questions of public versus private funding of the new plants, as well as bitter infighting among members for control of the committees' work. During the winter of 1939–40, the so-called "National Defense Power Committee" and then the "National Power Policy and Defense Committee" succeeded in conducting conferences with the representatives of electrical utilities from US war-production centres, but accomplished little else.[4] Wide public-private disagreement existed over expected load growth and thus the nature of a future power shortage. Furthermore, the government and the private sector could not agree on the feasibility, the funding, or even the necessity of a grand FPC plan to interconnect the major industrial centres of the Northeast by means of high-tension transmission lines in a giant power grid.[5]

Meanwhile, the war's outbreak in September 1939 produced a sharp rise in US electrical demand to meet anticipated growth in the manufacture of war materiel. Private utilities, in control of over 80 per cent of the total generating capacity of the nation, proved reluctant to publicly admit to a power shortage lest it mean further government encroachment upon the electrical business. Nevertheless, corporations responded with action, constructing new power plants and substations, transmission and distribution lines, stepping up voltages of existing lines, and protecting equipment from sabotage. By mid-1940, the industry organ *Electrical World* could report that private firms had slated over $1 billion for expenditure in 1941–43 to create new generating capacity of over 2 million kilowatts. This would form roughly one-half of new power capacity installed during the war in the United States to meet the burgeoning demands of defence production.[6]

The remaining half would come from government.[7] Hitler's successful invasion of Denmark, Norway, the Netherlands, and Belgium, and finally France, in the spring of 1940 shocked the still-isolationist United States into action on the industrial front. The Nazis' devastating coordination of ground forces with military air power showcased the airplane as a tool of war. Aluminum was the essential component of military aircraft, and electrical power was the controlling factor of aluminum production.

"When you are fighting with aluminum," an American power expert put it, "you are fighting with kilowatt[s]."[8] Roosevelt appeared before Congress on 16 May calling for an aircraft industry capable of manufacturing 50,000 military and naval planes.[9] Where power was concerned, the president attempted to clear the bureaucratic air by granting the Federal Power Commission, chaired by his old colleague from New York State Leland Olds, the authority to direct power planning for defence.[10] Olds would have to coordinate his efforts with the Power Unit[11] of the Advisory Commission to the Council of National Defense (NDAC)[12] that was headed by Charles Kellogg, president of the Edison Electric Institute. Olds, a New Deal Democrat, was a long-time advocate of public power and of government regulation of electrical utilities; Kellogg, a Republican, championed free enterprise. No wonder the two men thoroughly distrusted each other's actions and motives.[13] Moreover, Olds and Kellogg were forced to make decisions with an equally diverse set of players in the US government's wartime bureaucracy, including Edward Stettinius, Gano Dunn, William Batt, and, later, Julius Krug.[14]

Nevertheless, from the summer of 1940, this somewhat dysfunctional group moved to assure the nation of ample power for war production, and they did so with considerable resolve. "There is going to be a general increase in demands for power all over the country," explained Dunn, in July, to a subcommittee of the US Senate considering an appropriation to expand Tennessee Valley Authority's (TVA) power facilities for Alcoa's aluminum production. "We are in a dangerous situation in regards to the national defense."[15] This was but the first of a series of requests for funds to enlarge TVA's hydroelectric generation; all were eventually granted. The TVA, expanded by over 800,000 horsepower, would become a vital hub of national defence production during the war. At one point over 42,000 men and women worked in various defence industries in this inland region of the US South. The Columbia River basin of the Pacific Northwest, under the authority of the US Department of the Interior, would form a second production hub, and would be served by the public power of Bonneville (1938) and then Grand Coulee (1942) hydroelectric dams.[16]

The small border town of Massena, New York, near the St Lawrence River, promised to anchor a third production hub – specializing in electrometallurgical manufacture – but for its reliance on imported power. At the turn of the century, the Pittsburgh Reduction Company (Alcoa's forerunner) had erected an aluminum smelting plant at Massena, drawing at first a meager 21,000 horsepower[17] from the

Massena Power Canal, a small man-made offshoot of the St Lawrence River that emptied into the neighbouring Grasse River.[18] Alcoa planned to vastly expand its industrial operation at Massena once hydroelectric development proceeded at the international section of the St Lawrence itself, the so-called Long Sault. But the St Lawrence's development was stymied by a lack of coordinated legislation by Washington, Ottawa, New York, and Ontario. In need of power, Alcoa reached across the river to purchase 60,000 horsepower from a Canadian firm, the Cedar Rapids Manufacturing and Power Company, while also buying a large stake in the Canadian company itself.[19] By 1915, Alcoa was importing power from Cedar Rapids via a transmission line stretched across the St Lawrence. Thereafter, the aluminum company remained reliant on Canadian power for the operation of its Massena plant.[20]

As the Second World War got underway, Massena proved to be the missing piece in Alcoa's larger wartime power puzzle.[21] As of autumn 1940, Alcoa (then the sole producer of primary aluminum in the United States) sought some 200,000 kilowatts/268,000 horsepower of electrical power to expand aluminum production nationally, including in the TVA area, in the piedmont Carolinas, in the Pacific Northwest, and in upper New York State. With NDAC's Kellogg and FPC's Olds acting as brokers, most of the necessary electricity had been located, placed under contract, and routed to serve the aluminum company's growing reduction works. At Massena, however, Alcoa required an additional 50,000 kilowatts (67,000 horsepower), roughly the amount of its current imports from Cedar Rapids, to expand ingot smelting. Securing this power was beyond the influence of American utilities and US government planners alike. Although Cedar Rapids was a private company, exporting power required a Dominion government permit issued on an annual basis. No additional power would flow south across the St Lawrence without the Dominion's approval. Worse, wartime power shortages in Canada might well move the Dominion to shut down exports altogether to the Massena plant.[22]

Through the autumn and winter of 1940–41, Kellogg and his NDAC colleagues fretted over the Massena problem, which was compounded for Kellogg by the fact that his nemesis Olds had taken exclusive charge of the necessary international negotiations. Kellogg understood that Massena's Canadian power was "being actively negotiated through the Federal Power Commission with the authorities in Canada." He was also repeatedly assured by Olds that the FPC chairman could and would secure the electricity for the Massena plant, whether via Quebec's

Beauharnois hydroelectric station, via an additional diversion of water at Niagara Falls, or via a complicated power diversion drawing on the municipal electrical supply of Cornwall, Ontario.[23] Kellogg was joined by NDAC's power expert Gano Dunn, who described to Olds the "increasing demands on him for aluminum [which were] coming down to [the] matter of power [and the] question of more power at Massena." Dunn urged Olds to get the "quickest possible results." Various others in NDAC emphasized the "defense necessity of this power" and impressed on Olds the importance of the 50,000 kilowatts of power hoped for from Canada. Grenville Holden of NDAC's aluminum staff, an economist from Eastman Kodak Company, was incensed at this bottleneck in the US defence program. "Every effort should be made to get power ... at Massena," he wrote. "The first and most urgent task is to get the power." Holden also had a radical suggestion as to how the United States might gain leverage with Canada on the issue: withhold any further deliveries of aluminum to Britain "until this Massena power is obtained."[24]

The problem of power was related to another obstacle regarding the import of ingot. In 1937, Aluminum Company of Canada inherited the lower Saguenay River from its parent company, Alcoa, including the Chute-à-Caron plant and the undeveloped water rights at Shipshaw. As part of the arrangement, the Canadian firm contracted to provide Alcoa with a large quantity of aluminum ingot (19,300 metric tons per year) for three years, beginning in January 1938. In January 1941, that contract expired. And with power supply short in Canada and the British aluminum contracts therefore in potential jeopardy, the Canadian government had refused to issue permits for the continued export of aluminum to the United States. Alcan cut its deliveries to Alcoa in half, which in turn alarmed NDAC officials charged with securing US aluminum supply.[25]

The United States would likely not obtain further power or ingot imports for Alcoa while Olds retained responsibility for the task. Olds was probably playing politics with defence production. He was among several in the Roosevelt administration (including assistant secretary of state Adolf Berle, Jr and the president himself) who had for years sought the consummation of a grand St Lawrence Seaway public power project to be carried out in co-operation with the Canadian government.[26] By late 1940, these men anticipated a treaty agreement with Canada on the subject. They were artfully using Massena's continuing power shortage, along with the needs of defence production generally, to encourage political support for the enormous, expensive, and terribly controversial

hydroelectric and navigation scheme.[27] Seaway advocates understood
that to alleviate Massena's energy problem would remove an important
argument in favour of their beloved project. Olds, for one, would do
nothing to place the Seaway at political risk. Thus his effort on behalf
of Massena power was at best half-hearted.[28]

Kellogg and his NDAC colleagues suspected as much once the St
Lawrence treaty was signed in Ottawa in mid-March 1941.[29] Even if the
accord could survive the "full-dress battle" over ratification in the US
Senate that opponents expected[30] (and it would not), the hydroelectric
component would require at least four years' work, so Kellogg understood that the Long Sault's development would not resolve the industrial
power shortage in Upper New York State. Frustrated by the months
of promises and prevarications from the FPC chairman, NDAC's staff
now planned "some other line of approach" to the Canadian problem.
Rather than rely on Olds, Kellogg made direct contact with the Canadian
government's point man on electrical power, power controller Herbert
Symington. Kellogg arranged to meet with Symington in Montreal in
hopes of finding some means of resolving the power shortage in the
industrial Northeast United States.[31]

Fortunately for the Americans, Canada's power controller was already
sympathetic to their cause and had taken important steps in this regard.
To resolve a looming power shortage in eastern Canada and the north-
eastern United States, Symington viewed the completion of the Saguenay
River's hydroelectric development as his primary wartime mission.

Born in Sarnia and educated at the University of Toronto/Osgoode
Hall, Herbert ('Herb') James Symington moved to Manitoba during the
Wheat Boom years, where he became a prominent Winnipeg corporate
lawyer and businessman. Earning the admiration of the very rich, if notoriously reserved, financier Izaak Walton Killam, Symington was invited
to become general counsel and vice-president of Killam's influential
brokerage house Royal Securities. Through Royal Securities' investments
in pulp and paper and hydroelectric projects, Symington soon sat on the
boards of several power companies and railways. A Liberal, he believed
in Canadian autonomy within the British Empire and a strong, independent relationship for Canada with the United States; he also admired
and considered as a friend the Liberal lion of the West, editor John
W. Dafoe of the *Manitoba Free Press*. Symington's kindly, grandfatherly
appearance masked a steely disposition: as author Philip Smith has put
it, his "mild, round face and pince-nez belied his strength and astute-
ness." Capable and influential in his own right, Symington was also a

protege of sorts of the formidable businessman-turned-politician C.D. Howe, "the chief builder of modern industrial Canada" and "a great leader of men." The two had worked together in the western grain business in the 1920s. When Howe entered federal politics in 1935 as the minister of railways and canals and then minister of transport, he tapped Symington as a director of Canadian National Railways (CNR) and then as a founding director of CNR subsidiary Trans-Canada Airlines, the predecessor of Air Canada. In the summer of 1937, as Howe set out to personally demonstrate the practicality of transcontinental aviation via the first-ever, dawn-to-dusk, cross-Canada flight, he selected the loyal Symington as one of his two fellow passengers. Symington would serve as president of Trans-Canada between 1941 and 1947.[32]

When the Second World War broke out, the fifty-eight-year-old Symington joined the elite ranks of 'Howe's Boys' the seasoned business executives called on by the now minister of munitions and supply to manage Canada's burgeoning wartime economy. As to which facet of the economy Symington would oversee? He clearly knew something about electrical power, as well as about airplanes, and he understood the vital production link between the two. In August 1940, on Howe's recommendation, Symington was appointed wartime power controller with the Department of Munitions and Supply. In this position, he possessed the broad authority "to control throughout Canada the production, generation, transformation, transmission, distribution, supply, delivery, sale or use of power to ensure the best possible utilization thereof."[33] He immediately set about analyzing available power capacities and future power loads, especially in Quebec and Ontario where Canadian and British governments and private industry were locating power-intensive war industries. Given the industrial importance of these two provinces, and wary of possible federal-provincial jurisdictional conflict, Symington was also quick to confer with his provincial counterparts in Quebec's Public Service Board and Ontario's Hydro-Electric Commission. Quebec's power officials had certainly not waited for the appointment of a federal power controller to ascertain the province's wartime energy needs, so Symington did not have to look very far or hard to gather the necessary data.[34] Within weeks of his appointment, Symington understood that there would be a "very serious shortage of power in the winter of 1941–42, with an increasing shortage in the succeeding years."[35]

Symington also understood that the shortage would not be limited to Eastern Canada. He travelled with Howe to Washington in November 1940 where the two met with National Defense Advisory Commission executives William Knudsen, and later William Batt and Edward Stettinius. (NDAC became the Office of Production Management (OPM) in January 1942, and eventually the War Production Board (WPB).) The Battle of Britain had raged all summer and into the autumn. "At the time the Allies had been simply pounded to death in the air," Symington recalled. "It looked like an air war." Thus the American war planners wanted power. They wanted power for aluminum, and they expected that Canada could provide it. When Symington informed them of Canada's own coming power shortage, Washington's war executives were very disappointed. "They thought we had a great deal more power than we had," Canada's power controller related, "and we held meetings for some time looking into the necessities for the war picture." It was during these consultations that Symington recognized his most important wartime mission. "I came out of those meetings firmly convinced that if there was one thing Canada had to do to save this war it was to make aluminum."[36]

From Washington, Symington and Howe proceeded north to the finest location for aluminum manufacture anywhere on the continent: Quebec's Saguenay River Valley.[37] The Isle Maligne and Chute-à-Caron power plants, which they inspected, collectively possessed a generating capacity of 800,000 horsepower. These structures were complete and benefited from the impounded storage of Lac St-Jean, all within easy transmission distance of the Arvida aluminum works. The entire power-ingot system had access to both a railroad line and a tidewater fjord. Raw materials such as bauxite could be brought virtually to the site of power generation and the ingot could be exported by rail or ship for further processing, all at low transportation cost. The obvious complication at the Saguenay by the autumn of 1940 was that there was no power to spare. The electricity from the existing generating plants, which was available in large surpluses before rearmament, was now entirely absorbed in producing aluminum for the United Kingdom under several contracts dated 1939 and early 1940 between the British ministry of supply and Alcan.[38] But Howe and Symington understood that there was also a clear advantage to the Shipshaw scheme. The existing facilities for water storage, power generation, and alumina (aluminium oxide) production, as well as available design plans for a new Shipshaw plant, suggested the Saguenay as the very best opportunity to get power, and thus aluminum, in rapid time.[39] Shipshaw might be completed in just eighteen months, instead of the three years associated with a virgin project. "This was clearly a golden opportunity to get cheap power for war purposes," the visitors realized. "We came back from that trip ... unanimous that it had to be built," reported Symington. "Shipshaw was the thing to do ... There was not any question ... as to the war essentiality of the Shipshaw plant."[40]

Before turning to Alcan, Symington and Howe gave at least brief thought to the government investing in new power plants. Howe considered such intervention unwise, however. In Canada, by contrast with the United States (where the federal government had intervened during the Progressive Era to restrict construction of, and eventually build its own, power projects[41]), power resources were owned and/or regulated by the provinces. Any attempt by Ottawa to intervene "would create a very uncomfortable position."[42] Moreover, Howe, the American-born cabinet minister who himself relished power, was nevertheless an ardent advocate of free enterprise and of self-interested free will.[43] Howe feared that direct government involvement in electrical power (or outright nationalization, for that matter) would remove the all-important ingredient of incentive from wartime industrial work. "The Federal Government will not enter the power field," Howe decided. Instead, "war work must be placed with full responsibility as to sources of power in the hands of the manufacturer." Symington concurred. He recognized that had Ottawa attempted to appropriate the Saguenay's hydraulic resources from Alcan, or to appropriate the corporation itself, there would certainly have been "tremendous opposition and furor" within Quebec. The result would have been a serious delay in industrial production. The war emergency was no time "to settle fundamental questions" of private versus public ownership, Symington felt, and the method adopted by Canada was by far "the better" for the outcome of the war.[44]

The dam-builder, then, would be Alcan – that is, if Symington could coax the aluminum company to take up the task. Alcan's executives had good reason to feel reluctant. The Great Depression, after all, was still fresh in memory. Between 1931 and 1937, the aluminum market was in collapse. Electrical power from the existing Isle Maligne and Chute-à-Caron plants had run largely to waste, while Alcan's top salesmen scoured the continent in an effort to entice industrial customers to the outland region. "We had surplus power," Alcan's Powell recalled of the 1930s. Said Dubose: "those were very trying financial years."[45] By late 1940, market conditions had, of course, vastly improved. Rearmament and then the war itself had absorbed the Depression's power surplus and also necessitated incremental expansion of power and smelting capacity.[46]

Still, no one, either in the company or in the government, could visualize sustained aluminum demand in the immediate *post-war* era at a level that would justify the expenditure of $30 million (at minimum) on a Shipshaw power plant. Such expansion was "neither foreseen nor hoped for," Edward Davis told shareholders after war's end. "Never in our wildest moment" of optimism, stated Powell in 1943, "did we imagine that we would be called upon to expand as rapidly as we did." Alcan director Earl Blough echoed those thoughts the following year: "No one, in his wildest flight of fancy, could have foreseen the [wartime] demands that were to be made on ourselves and the other producers."[47] Alcan had already multiplied its aluminum-production capacity five times since rearmament began. By war's peak (1944), that multiple was ten: Alcan possessed the electrical power and manufacturing facilities to produce 500,000 metric tons of primary aluminum each year, an amount that was equivalent to what the entire world had made during 1937![48] Shipshaw was likely to be "wholly superfluous at the end of the war," thought Frank Brown, the Department of Munitions and Supply's financial adviser. "It will be idle or very little used for a number of years." Symington himself believed that Shipshaw was a "gamble" given the uncertainties of post-war demand and that the ingot produced as a result of Shipshaw's huge energy "would be largely excess."[49] (None of these men could have foreseen the rapid increase in peacetime use of aluminum in the post-war period, let alone the demands of the Cold War, both of which would require Alcan, with the outbreak of the Korean War in 1950, to further expand its energy and smelting facilities at the Saguenay and elsewhere.[50])

Whether Shipshaw proved superfluous or not, Symington was determined to get the dam built. During the winter months of 1940–41, Canada's power controller sweated over the problem of Canadian aluminum production if Shipshaw were not developed.[51] "The result of this war depends largely on the air development and the air development depends largely on aluminum," Symington wrote Howe. "I am of the opinion that somebody is going to wake up suddenly and demand aluminum which means a demand for power, and power will not be available." Symington also bluntly told the minister that he "would no longer take responsibility for the power situation unless Shipshaw was developed." The power controller was equally forceful with Alcan's executives. "I went after the Aluminum Company just as hard as was possible," he recalled of the period November 1940 through March 1941. "I was lashing them all the time to get it done." Yet the company resisted, insisting that its shareholders should not be called on to fund what would likely turn out to be an idle plant. Said Symington of Alcan: "they cried like a wounded moose."[52] By late February, the power administrator was losing patience. "I am putting it squarely before the Aluminum Company of Canada," he wrote Alcan president Powell, "that the projected power development at Shipshaw should be started immediately and completed as speedily as possible."[53] Still, matters dragged on for one more month. Late in March, Alcan finally relented, agreeing to build at least a partial development of Shipshaw (500,000 horsepower of the possible one million) at its own expense, as long as the Dominion promised to provide significant financial aid in the form of tax relief.

As Symington struggled to alleviate Canada's power shortage, his boss Howe, minister of munitions and supply, was trying to resolve Canada's equally consequential shortage of US dollars. Canada's wartime purchases in the United States of defence equipment destined for Great Britain had produced, by the early months of 1941, a serious imbalance in Canadian-American currency exchange and, for Canada, a growing scarcity of US dollars. Canada's financial experts in the Bank of Canada and Department of Finance had wrestled with the problem from the war's outset, introducing measures to conserve foreign exchange. In March, the US Congress passed the Lend-Lease Act; its purpose was large-scale assistance to Britain and the Allies, but it, nevertheless, threatened to divert British war orders from Canada to the United States and exacerbate the crisis. Howe told his colleagues in the War Cabinet that he was "gravely concerned" with this possible outcome. Financially, Canada now risked facing what Prime Minister Mackenzie King called "a greater burden than [its] people ... can be led to bear." At stake, quite simply, was Canada's ability to continue to fund its war effort. Staff in the Department of Finance generally opposed seeking help from the United States by including Canada in the list of Lend-Lease nations; in the long run, this stood to weaken Canada's trade position vis-à-vis its southern neighbor.[54]

In early April, Howe added his influential voice to those opposing Canada's participation in Lend-Lease. Howe instead suggested that the United States co-operate with Canada by purchasing certain war materials that Canada was already producing and that were needed south of the border. Aware of Alcan's pending expansion at the Saguenay, Howe also suggested that Canada could increase its production of raw aluminum, which would supply a much-needed item to the United States and, simultaneously, provide Canada with badly needed US dollars.

Finance's deputy minister Clifford Clark as well as Prime Minister King endorsed Howe's strategy. It was also embraced by US secretary of the treasury Henry Morgenthau, Jr when King met with him in Washington on 17 April. Morgenthau obtained President Roosevelt's approval for the cross-border plan before working out its details with the Canadians.[55] The Hyde Park Declaration, which was announced by the president and prime minister on 20 April at Roosevelt's Hudson River estate, embodied the general principle of collaborative cross-border trade that Howe had suggested two weeks earlier. "In mobilizing the resources of this continent," stated the Hyde Park Declaration, "each country should provide the other with the defense articles which it is best able to produce, and, above all, produce quickly, and production programmes should be coordinated to this end."[56] Howe heaped praise on the prime minister upon King's report to the War Committee on what had transpired at Roosevelt's estate.[57]

In this study of the Saguenay basin, it is useful to highlight the important presence of "aluminum" in the far-reaching agreement that was made along the Hudson River. Howe could recommend Canadian aluminum exports because he had toured the Saguenay the previous autumn with Symington and persistently heard from the power controller over the preceding several months about the need to build Shipshaw. Howe passed on his recommendations to Mackenzie King, who, in turn, specifically broached the subject of aluminum in his meeting with Morgenthau on 17 April. Aluminum was an excellent carrot to offer the United States. By the last weeks of 1940, US war-production executives were stung by serious shortages of aluminum that were curtailing aircraft production by such major firms as Northrop Aircraft Incorporated and Glenn L. Martin Company. Come mid-May, Senator Harry Truman, in his hard-driving investigation of these aluminum shortages, would be harshly criticizing those responsible at Alcoa and in the US government. Aluminum had become a political, as well as a war-preparedness, issue in the United States.[58] No wonder Morgenthau, in his meeting with King, noted the "tremendous help" it would be to get Canadian aluminum; and with Clark and other Canadians on 18 April, had asked "And what about aluminum?" to ensure that the light metal be specifically added to the list of materials that Canada aimed to supply. Again, when phoned by President Roosevelt from Hyde Park, Morgenthau questioned whether "aluminum" was articulated in the planned press release before recommending to the president and prime minister its immediate distribution. (According to King's memory of events, it was Roosevelt himself who added aluminum to the draft agreement.) One day after the summit, Clark submitted to Morgenthau a full list of articles that Canada could supply. The largest single item on the list for the first year of planned US purchases was "Metals and Minerals," coming in at $66.7 million; and by far the largest component of this section, valued at $30 million immediately to the Canadians, was the pending contract between Aluminum Company of Canada and the US government. Howe understood this contract to represent "a forward step in implementing the Hyde Park Declaration." Its passage, recalled Munitions and Supply's financial adviser Frank Brown, produced "a good deal of feeling of relief" among the staff of Canada's Department of Finance.[59]

Aluminum ingot that was to be manufactured at the Saguenay River was an important catalytic agent for the Hyde Park agreement, which may help to explain why the Americans so readily agreed to the Declaration's terms.[60] Similarly, Hyde Park provided a powerful endorsement, a mandate of sorts, for Alcan's giant ingot contracts with the US government. The contracts themselves seem to have originated in the renewal by Alcan of ingot contracts with Alcoa, which were discussed at the end of March 1941 in Montreal. Within a week, Washington's war planners decided that the US government's Metals Reserve Company should replace Alcoa as the purchaser of Canadian ingot.[61] Then, in the wake of Hyde Park, Alcan's Powell and MacDowell and Metals Reserve Company executives W.L. Clayton and S.D. Strauss hastily negotiated the first contract in Washington.

Alcan was the reluctant party in these negotiations and played this situation to good advantage. It was, after all, "representatives of the Canadian Government," Powell later explained, that "urged" Alcan to make its agreements with the US government.[62] Since Alcan refused to sell ingot to the Americans without advance funding for the Shipshaw-Passe Dangereuse project, such advances were built into the contracts. Powell and MacDowell also drove a hard bargain on the ingot's price. Under the contract of 2 May 1941, Alcan undertook to deliver 375 million pounds of ingot at a base price of 17 cents per pound (somewhat above prevailing market rate, but later reduced by mutual agreement to 15 cents). An 'escalator clause' allowed the price to increase in response to rising wartime costs of power, labour, or transportation. Metals Reserve advanced $25 million as a down payment against the metal in order to fund the construction of the necessary power and production facilities. In July, under pressure from US war planners, Metals Reserve doubled its order. This second contract was written along similar lines, with Metals Reserve advancing an additional $25 million to Alcan. This latter contract also prompted the Canadian company – again under heavy pressure from the power controller – to more than double Shipshaw's planned generating capacity, from 500,000 to 1.2 million horsepower.

The larger development necessitated water storage: it was simply not possible to run water through all twelve of Shipshaw's planned turbines for 365 days each year without the additional regulation of the Saguenay's flow at the Peribonka River. In this way the Passe Dangereuse reservoir became an essential component of the larger hydroelectric scheme. Lots of power also promised large quantities of aluminum. Metals Reserve ultimately received over 1.3 billion pounds of ingot from Alcan during the wartime period, which amounted to roughly one-fifth or 21 per cent of the primary aluminum that was produced in the United States itself by both the US government and private industry.[63] Metals Reserve paid Alcan roughly $250 million for this ingot during the course of the war. This included the advanced payments, which comprised virtually the entire $70 million necessary for the construction of the Shipshaw and Passe Dangereuse dams. This fact would later generate considerable controversy in the United States.[64]

Of course, the Americans' most significant concession didn't concern the expenditure of US dollars per se, but rather the location of the dams on Canadian soil. Both politically and strategically, Washington's war planners would have much preferred to import additional Canadian power to Massena, rather than build new power dams outside the boundaries and jurisdiction of the United States. In conceding to fund Shipshaw and its related storage works, Metals Reserve's executives secured a concession of their own: if the Americans bought ingot from Canada made at the Saguenay plant, the Canadians would keep the power flowing to New York State.

Washington required that Symington guarantee the continued flow of electrical power across the St Lawrence River to Massena, and Symington complied. The idea was that once Shipshaw was generating electricity, the Saguenay valley's aluminum works would function independently, that is, without additional transmission of power from the surrounding river basins of Quebec. In the early war years, Powell explained, the Arvida works "had been draining the whole of Eastern Canada for power."[65] With Shipshaw built, Canada's power controller could then release Canadian power from elsewhere in Quebec or Ontario to serve upper New York State. The American war administrators (as well as

Alcoa's executives) hoped that, in this way, Shipshaw would yield an additional 70,000 kilowatts (some 94,000 horsepower) of firm power for Alcoa's use at Massena for the duration of the war, and they pushed this as a condition of the original ingot contract.[66] Symington took a more conservative position, promising only that Shipshaw's completion would allow the Canadians to continue exports to Massena at the current level – 108,000 kilowatts or 140,000 horsepower – with additional exports but "a possibility ... depending ... on the situation as it should then be."[67]

The Canadians were also determined that the matter be handled quietly, outside diplomatic channels, lest the sensitive issue of power exports to the United States be exposed in the Canadian Parliament. Howe laid out the matter for the War Committee of the Cabinet. The Cabinet recommended that Symington be authorized to offer assurance to the Americans in this regard. Canada's power controller then wrote a letter to the Metals Reserve Company in which he guaranteed that the present power exports to Massena would be continued (albeit for the duration of the contract between Metals Reserve and Alcan), "and if conditions permit ... increased." This letter became part and parcel of the package of documents that became the first Alcan–Metals Reserve contract of 2 May 1941. It was signed on 11 June, once this and several other conditions had been worked out between Metals Reserve, Aluminum Company, and the Dominion.[68]

Metals Reserve assistant vice-president Strauss summarized the sequence of events as follows: "On the basis of the [ingot] contract, the Aluminum Company of Canada proceeded with the power expansion [at Shipshaw] which made possible the commitment on Massena."[69] We should note that Canada honoured this commitment to continue power exports to the Massena plant through the peak war-production year of 1944. However, the evidence suggests that the additional power sought so ardently by the Americans over the course of several months was never delivered. The Americans replaced it with expensive steam power that was generated by Consolidated Edison Company of New York City and transmitted north at considerable cost. Nor did that downstate electricity serve an Alcoa plant, as originally intended. Massena, New York, became one of several sites nationally where the US government funded its own aluminum plants for war production, drawing on the construction and managerial expertise of the private sector. The so-called St Lawrence aluminum reduction plant (originally 'Plancor Plant') was built and managed by Alcoa for the US government's Defense Plant Corporation, adjacent to the Massena Works, and then purchased by Alcoa from the government some years after the war.[70]

Since Alcan also refused to undertake expansion without substantial tax relief from the Canadian government, further concessions were forthcoming on that front. Again, Howe took the lead. On the train coming down from the Saguenay in November 1940, Howe decided to allow Aluminum to write off 50 per cent of construction costs in order to coax the firm to proceed with Shipshaw and Passe Dangereuse. The idea was that these developments served wartime demand exclusively, and their post-war values would plummet (or rapidly 'depreciate'). In Howe's estimation, the value would fall by roughly one-half. When Howe made the offer to Powell, Powell pressed for a figure of 100 per cent, arguing that the developments would be worthless. Howe would give Powell verbal assurance of a 'special depreciation' rate of 60 per cent, as long as Aluminum agreed to build the entire 1-million-horsepower Shipshaw development, which necessitated the storage contribution of a Passe Dangereuse reservoir. Howe directed Powell "to go ahead [on the dams] and we will get this thing fixed up." Howe also suggested that, since one of its aims was to acquire US dollars, the agreement could be legally resolved under the War Exchange Conservation Act. Powell saw the Department of Finance's Clark to firm up an agreement on paper.[71]

Thus began lengthy and complicated negotiations across some twenty months – from the spring of 1941 through December 1942 – by which the government of Canada sought to work out a method of tax relief that was politically defensible as well as financially amenable to Aluminum Company. Howe's verbal assurances aside, being forced to commence construction of dams and factories long before an agreement became law caused Aluminum Company executives considerable anxiety. Howe, as well as Symington, was kept fully informed of the ongoing negotiations by staff in the Department of Finance. Also kept advised was Godbout's minister of the treasury, English-speaking Montreal attorney James Arthur Mathewson, with whom Aluminum advocate Geoffrion worked out the details of Aluminum's financing. The specific terms of the agreement posed only half of the difficulties and frustrations experienced by the negotiating parties. In addition, it became apparent by early 1942 (once Great Britain contracted for more ingot and the United States entered the war) that the matter could no longer be legally resolved under the jurisdiction of the War Exchange Conservation Act and its Control Board. It was turned over to Justice C.P. McTague of the War Contracts Depreciation Board (and the Ontario Supreme Court). McTague, too, passed on the task, judging it outside his own board's jurisdiction for a variety of reasons. But it would be McTague's recommendations that formed the body of the agreement. In the end, the accord was granted by a special Order-in-Council under the War Measures Act. It gave Alcan permission to write off 60 per cent of its capital investment in all power facilities, roughly $41 million, during the course of the war, as well as the entire value (some $120 million) of related expansion in mining, manufacturing, and transportation facilities.[72]

This digression from the provincial stage to the continental one, from Quebec City to Ottawa and Washington, yields at least three relevant conclusions.[73] First, the Passe Dangereuse reservoir was essential to the Saguenay's hydroelectric expansion if Alcan was to build Shipshaw to its full capacity. Therefore, the Peribonka reservoir was linked hydraulically and technically, as well as financially and politically, to a giant industrial project that held thoroughly North American and, indeed, global implications. Second, as a result, the entire hydro project had powerful supporters. Both Symington and Howe had shepherded Shipshaw as their own from the autumn months of 1940. Shipshaw-Passe Dangereuse served Canada's (and the northeast region's) urgent needs for electrical power, as well as the nation's critical needs for US dollars. As the work and conclusions of Canada's War Expenditures Committee would later suggest, it also served Canadians' patriotic desires to fight the Axis by developing the nation's industrial potential.

In all these respects, third and finally, Godbout had excellent cause to dispense with the formalities of provincial law in approving dam construction prior to formal government authorization. The premier very likely understood the regional, national, and continental ramifications of the Saguenay's development for the world war as early as the winter of 1940–41, when Shipshaw was conceived. If he did not, the power officials of Quebec's Public Service Board could certainly have explained it to him.[74] So, for that matter, could have Prime Minister Mackenzie King. King sent a lengthy letter to Godbout's minister of lands and forests Côté in November 1941 describing the "great aluminum expansion program now under way at Arvida, P.Q." Explained King (clearly having been briefed by Howe and/or Symington): the United States government's Metals Reserve Company had contracted to buy vast quantities of aluminum from Aluminum Company of Canada for the production of war materiel. In order to serve these contracts, Aluminum Company "must very substantially increase its plant facilities," for which the US government had advanced the money.

> By reason of the willingness of the United States Government
> to advance funds ... this Company was enabled to proceed with
> the immediate construction of a new plant – thus Quebec is
> benefitting under these arrangements by this great industrial
> development within its boundaries. In advancing these funds,
> the United States Government asked, in return, for a guarantee
> that the export of power to the Aluminum plant at Massena
> would not be curtailed during the life of these contracts and
> would be increased if conditions permit. This guarantee was
> given by the Power Controller to Metals Reserve with the
> approval of the Canadian Government. While there is no ques-
> tion at present of increasing supplies of power to Massena, and
> conditions which might permit of increase are not in sight, it is
> certain that these supplies must not be curtailed.[75]

Côté shared a copy of the letter with Premier Godbout and discussed
the matter personally with him.[76]

Godbout certainly felt and understood the weighty momentum of an
industrial juggernaut that derived from Washington's calls for power and
ingot, Ottawa's needs for foreign exchange, and the Canadian prime
minister's desire to maintain in good faith the Canadian-American wartime military-economic alliance that had been sealed at Ogdensburg,
NY (August 1940) and then Hyde Park. It was highly unlikely, then, that
Godbout's administration, already supportive of Canada's war effort,
would have done anything to delay a project of such international magnitude and importance.

If Godbout showed any reluctance to break with the established
rule of law, he would have been quietly urged to do so by powerful
Liberals in the orbit of the King administration – be it Howe; King's
French-Canadian lieutenant, Erneste Lapointe; or any number of
intermediaries – or Mackenzie King himself, who held Godbout in the
highest esteem.[77] The trick for Ottawa's men was simply to avoid publicly
offending provincial jurisdictional sensibilities. After all, there was (and
still is) no greater danger for a Quebec politician than to be perceived
as doing Ottawa's bidding. The Opposition's vicious attacks throughout
the war years that Godbout was doing just that, would bring about not
only the downfall of the Godbout era but also a half-century's ostracism
of his name from Quebecers' collective memory: not a dam, not a street,
not a public work named for him or erected in his honour until the
year 2000.[78] Symington was the supreme authority on matters of power

production during the war years. He was determined to do his utmost to make the province of Quebec a Canadian arsenal for the production of aluminum, yet he was also sensitive to provincial jurisdiction over natural resources.[79] If Symington wished to communicate his wishes discreetly, he could do so via the English-speaking member of Quebec's Public Service Board, engineer John McCammon (who, like Latreille, would go on to become one of the founding commissioners of Hydro-Québec), with whom he was in close contact during the war. It is doubtful that Symington dealt directly with Godbout. Symington doesn't seem to have thought highly of "Frenchmen" generally. By the late fall of 1943, he found Quebec's premier to be "very inept" as a Liberal leader; and he deemed Godbout's intention to nationalize Montreal Light, Heat and Power "a dangerous thing, and one very doubtful whether it will result in benefit to the public involved."[80]

Premier Godbout's informal approval of Passe Dangereuse would likely have been accompanied by a stern warning to Aluminum Company to complete the negotiations with the Hydraulic Service or risk losing possession of the dam. Given the premier's penchant to delegate, rather than dominate, decision-making, we can also speculate that Godbout discussed the political ramifications of such a decision with his colleague Côté, who, in turn, would have briefed the Hydraulic Service's Latreille. All such highly sensitive communication would, of course, have taken place by telephone, or face to face, leaving behind no trail of paper.

C.D. Howe and Herbert Symington, New York, 1941. Canada's power controller Herbert Symington (right), long-time business partner and friend of war production czar C.D. Howe, discovered his primary wartime mission in the United States: mobilizing Canada's power resources in order to smelt aluminum for Allied aircraft. "If there was one thing Canada had to do to save this war it was to make aluminum." With Howe's backing, Symington targeted the huge Saguenay River for further hydroelectric development and pressed Aluminum Company of Canada to build the necessary dams. Howe and Symington were photographed at La Guardia Field (LaGuardia Airport) in Queens, New York, likely as they inaugurated Trans-Canada Airlines' Toronto to New York City route. (Public Archives of Canada, PA-092483)

J.B. Duke and associates, 1915. It was American businessmen who
spearheaded the Saguenay River's hydroelectric development and
envisioned its watershed-wide exploitation. Principal among these was
North Carolina tobacco magnate James B. ('Buck') Duke (third from
right). In 1915, intending to use electricity to burn nitrogen from the air
and manufacture nitric acid for gunpowder, Duke showed his Canadian
power properties to executives of the Wilmington, Delaware–based E.I. du
Pont de Nemours Powder Company (Du Pont). Left to right: local agent
Benjamin Scott; Du Pont's Linden Edgar, Lammot du Pont, and Morris
Richard Poucher; Duke; business advisor George Allen; engineer William
States Lee. Duke eventually merged his interests with Aluminum Company
of America in 1925, which, three years later, spawned a Canadian subsidiary
that became Aluminum Company of Canada. (Rio Tinto Alcan, Montreal,
ALBU-15-01)

Ray Powell. Aluminum Company of Canada president Ray ('Rip') Powell, who oversaw the firm's rapid global expansion during the Second World War, had peddled aluminum cookware in his native Illinois before rising through the ranks of Aluminum Company of America's sales staff. (Rio Tinto Alcan, Montreal)

Ray Powell with Canada's governor general and the bishop of Chicoutimi, 1942. In recognition of Aluminum Company of Canada's contributions to the war effort, Powell was commissioned honourary lieutenant colonel by the governor general, the Earl of Athlone, as the bishop of Chicoutimi looked on. (Rio Tinto Alcan, Montreal, ALBU-42-02)

McNeely Dubose. Power expert and Aluminum Company of Canada vice-president McNeely ('Mac') Dubose, originally from North Carolina, supervised the construction of hydro plants in Quebec. Dubbed 'king of the Saguenay' by his colleagues, Dubose also took the lead in negotiating concessions from the Quebec Government for the creation of lakes Manouan and Passe Dangereuse. (Rio Tinto Alcan, Montreal)

Portrait of Maurice Duplessis, 1936. Premier Maurice Duplessis is oft recalled as a champion of big business, but he rejected Aluminum Company of Canada's demands to impound Lac Manouan. (Archives du Séminaire de Trois-Rivières)

Adélard Godbout, 1939. Duplessis's more progressive successor Adélard Godbout, seen here in the campaign of 1939, favoured Aluminum Company of Canada with wartime concessions of both Lac Manouan and Lac Passe Dangereuse in order to serve Canada's war effort. (Collection famille Godbout/ La Commission de la capitale nationale du Québec)

Godbout and Churchill, 1943. Premier Godbout greeted Winston Churchill at the Quebec City Conference in August 1943. Godbout's concessions to Aluminum Company of Canada drew vicious partisan attacks from the Opposition, which, in turn, seem to have forced the premier's decision to create Hydro-Québec. Godbout's wartime co-operation with Ottawa would cost him re-election as well as his political career and reputation. (Collection famille Godbout/La Commission de la capitale nationale du Québec)

Portrait of Raymond Latreille. As chief of the Hydraulic Service between 1940 and 1944, civil engineer and proud hydro-nationalist Raymond Latreille managed Quebec's waterpowers, including their concession to private interests. Latreille negotiated with Aluminum Company of Canada to achieve the best possible lease terms for the Province in the Peribonka valley while refusing the company access to the neighbouring basins of the Bersimis, Manicouagan, and aux Outardes rivers. (Archives d'Hydro-Québec)

Commissioners of Hydro-Québec taking possession of Montreal Light, Heat and Power head offices, April 1944. Latreille became one of the original and youngest commissioners of Hydro-Québec. Left to right: president T.D. Bouchard; commissioners George McDonald, Raymond Latreille, L.E. Potvin, and John McCammon. (Archives d'Hydro-Québec)

Latreille presents Manic 5 at press conference, 1960. Later in his career, Latreille explained Hydro-Québec's Manicouagan-Outardes hydroelectric project, including the massive multiple-arch Manic 5 dam, to members of the press. Latreille was the only member of the original Commission to oversee full-scale nationalization of Quebec's electricity industry in 1963. (Archives d'Hydro-Québec)

Pointe-Bleue, 1907. Innu families spent several weeks each summer at reserves camping in tents or residing in cabins. (Société d'histoire et d'archéologie de Mashteuiatsh, no. 5)

Joseph Dominique and family, 1945. Pointe-Bleue hunter Joseph Dominique and his family in front of their tent. (Société d'histoire et d'archéologie de Mashteuiatsh, no. 158)

Bersimis, ca. 1920s. This photo, taken from the church belfry, shows
hunters' cabins built along the estuary of the St Lawrence River. Such homes
would have been unoccupied through the fall, winter, and spring seasons.
(Collection Uashkaikan, RU16)

Poling upstream, ca. 1900. Families ascended to their hunting territories after
the Catholic *Fête* of mid-August. At the time this photo was taken, birchbark
canoes were giving way to wood-and-canvas crafts; both types are seen here.
(Société historique du Saguenay, Fph 65.1070)

Bersimis hunters pole upriver, late 1930s. As each canoe bore a load of 700 to 900 pounds, negotiating quick water by use of the *perche* required strength as well as skill. (BANQQ, Collection Paul Provencher, Neg. 135)

Sylvestre Kapu portaging. Sylvestre Kapu portages sacks of flour and a pail of lard. A bark head protector distributes the strain of the tumpline. (BANQQ, Collection Paul Provencher, Neg. 215)

Sylvestre Kapu and family. Sylvestre Kapu and his family portage upriver. (BANQQ, Collection Paul Provencher, Neg. 147)

Departing by car from Pointe-Bleue. During the interwar period, Pointe-Bleue
Natives began using hired cars or trucks to skirt the windy expanse of Lac
St-Jean. (Société d'histoire et d'archéologie de Mashteuiatsh, no. 837)

Family of Simon Raphael, ca. 1936. Through the 1930s, the vast majority of
Innu families – multigenerational groups that included men, women, and
children – spent the fall, winter, and spring seasons in the bush. (Société
d'histoire et d'archéologie de Mashteuiatsh, no. 329)

Fox pelt negotiations, 1930s. The Innus' continued reliance on forest materials is evident in this and the following two photos taken by French-Canadian forester Paul Provencher at winter camps along the Manicouagan River. Here, Adhémar Martin and Malec Collard negotiate purchase of a fox pelt with trapper Felix Poitras in front of a canvas round-tent. (BANQQ, Collection Paul Provencher, Neg. 100)

Joseph Desterres. Joseph Desterres of Bersimis uses the traditional crossknife to carve a snow shovel. Two granddaughters are with him in front of his cabin. (BANQQ, Collection Paul Provencher, Neg. 625B)

Madame Joseph Vachon. Madame Joseph Vachon, wearing a Christian cross about her neck, completes the lacing of a snowshoe. (BANQQ, Collection Paul Provencher, Neg. 636)

Jack Germain, 1979. Jack Germain, band chief and Peribonka valley hunter, guided Aluminum Company of Canada survey crews to the sites of both the Lac Manouan and Passe Dangereuse reservoirs. Later in life, he came to regret the role he and others played in the demise of a hunting society. (Société d'histoire et d'archéologie de Mashteuiatsh, no. 396)

5

"Do We Live in the Province of Aluminum or in the Province of Quebec?"

The political fallout materialized as dam construction neared completion. Interestingly, it was Shipshaw, rather than Passe Dangereuse, that first ignited the controversy. The 'Shipshaw Scandal' broke in the United States in the spring of 1943, after *The New York Times* reported in January on the completion of the Canadian dam, noting that the US government had essentially paid for it with its purchases of ingot. Roosevelt's secretary of the interior and public power advocate Harold Ickes felt certain that Shipshaw's enormous energy threatened the US public power movement. Residents of the Pacific Northwest and elsewhere, led by Representative John Coffee, decried the Canadian project as a "Frankenstein monster," assembled and brought to life with US dollars but threatening to destroy American jobs and industry.[1] This American controversy spilled over the border into Canada after M.J. Coldwell, leader of the Co-operative Commonwealth Federation and Canada's own champion of public power, met with Representative Coffee in the United States. Coldwell returned to Canada to accuse the King administration of favouring Canada's aluminum monopoly, Aluminum Company of Canada (which Coldwell repeatedly called "the creature" of Alcoa), by offering the firm a 'gift' of special tax breaks on Shipshaw's construction. Coldwell charged Alcan with wartime profiteering and blamed the King government for failing to bring the Saguenay's critical and enormous energy resources into the public domain. "I say the united nations are being sold down the Saguenay River by these agreements," he stated.

These accusations resulted in the appointment of a Special Committee of parliamentarians to inquire into Canada's war expenditures. By early 1944, following forty-two seatings and a tour of Alcan's factories at Arvida and Kingston, the committee had exonerated Aluminum Company and the Canadian government of all wrongdoing

(Coldwell dissented). The committee members (Coldwell excepted) also warmly thanked and applauded Powell and Alcan for a "record of ... achievement ... [of which] all Canadians can be proud," agreeing with Symington's assessment that Alcan's expanded power and plant facilities "constituted Canada's greatest contribution to the war effort."[2]

The provincial Opposition also took inspiration from the controversy unfolding south of the border. Given that the control of hydro resources in Quebec was historically a sensitive issue, and given that Godbout had waived provincial law, it was quite possible that Opposition criticism would find more traction at the provincial level than it did in Ottawa.

Throughout the legislative session of 1943, Duplessis seized on the transgressions of the Godbout administration to harass the government for pandering to the electrical "trust." "Hydraulic powers were bequeathed to a people to ensure their ethnic survival," said Duplessis, sounding his main theme. "But our immense natural resources, in particular our hydraulic forces, which rank third in the entire world, have been sacrificed by Liberal administrations for a mouthful of bread." Duplessis claimed that he, as premier, had "rejected [Aluminum's] request [for the Peribonka], saying that never would the province of Quebec become the province of aluminum [and] believing that the Peribonka power was ideal for a state co-operative development."[3] "But after the election of 1939, the Company got what it wanted. With contempt for all the laws of the province, without depositing the plans and drawings necessary, without first obtaining authorization, the Aluminum Company illegally commenced to build the dam [while also] ravaging our most beautiful forests." In this way did the administration "prevent for years to come any possibility of nationalization, by granting control to [Aluminum] of a million horsepower."[4] Duplessis charged that "the Peribonka affair is a scandalous affair ... The Aluminum Corporation has become queen and master of the Saguenay-Lac St-Jean region ... When the present government goes to the people, the latter will give it the proper answer, a definite and unmistakable condemnation."[5]

That Duplessis was conflating the Peribonka, Manouan, and Saguenay projects in his various remarks was, by itself, politically manipulative. More egregious and hypocritical was the Opposition leader's purported support for the nationalization of electricity in Quebec, when in fact he had long and vehemently opposed any such thing. Nevertheless, the charge of breaking provincial law was accurate, as was the accusation of drowned timber, and Hamel and Godbout were forced to respond. As the charge of law-breaking was undeniable, Hamel and Godbout

avoided it altogether, to the derision of Duplessis. Instead they emphasized, as Liberal politicians had done for decades, how foreign industry's
large investments in waterpower and factories served to employ Quebec
workers, to industrialize the province, and to raise the living standard
of its inhabitants. (As Taschereau was wont to say in the 1920s, "It is
better to import American dollars than to export French-Canadians to
the United States.") Hamel called attention to the fine revenues that
would accrue to government from the concession of the reservoir, to
the compensation fisheries received for the first time, and to the fair
compensation forthcoming for the timber. "It would certainly have been
preferable to cut one hundred percent of the forest," said Hamel, "but
the necessities of war prevented this."[6] Hamel pointed out that the new
reservoirs had helped mitigate the provincial electricity shortage, in turn
permitting Quebec to arrange with the Federal power controller for the
province's exemption from power restrictions on the paper industry.
Quebec was the only Canadian province with such a privilege of exemption, he stated.[7] Hamel also touted the benefits of flood control that
accrued from both Lac Manouan and Passe Dangereuse. Just that spring
(of 1943), Hamel told Quebec journalists, despite unusually deep snows
and a rapid thaw, the reservoirs' enormous absorptive capacity "saved
the district's population from floods analogous to those of 1928."[8]

These were compelling counter-arguments. Still, Duplessis's
remarks touched a nerve among those who had long sought greater
French-Canadian influence over the province's hydraulic resources.
The ardent nationalist and Independent member of the Legislative
Assembly René Chaloult remarked, "I do not doubt the honesty or
patriotism of the Premier. I note only that he isn't any more able than
the Union Nationale to free us from the yoke of the foreign trusts ...
This Government had promised to nationalize electricity ... I despair of
ever seeing the nationalization of our natural resources in the province
of Quebec."[9]

Back in 1939, Chaloult, like other former and disaffected members of
the Union Nationale, including Oscar Drouin and Philippe Hamel, had
set his hopes for progressive reform on the agronomist premier and his
new Liberal government. Similarly, Godbout had courted the nationalist
leader, granting Chaloult essential political support in the election campaign and inviting him to be a member of the Liberal caucus. By this point,
however, the relationship had soured, as had the temporary and fragile
coalition between the Liberals and nationalists. Chaloult and his colleagues viewed Godbout's wartime collaboration with Ottawa – regarding

a program of unemployment insurance, jurisdiction over taxation, and most significantly, the imposition of conscription – as a clear 'capitulation' to King, and a clear betrayal of provincial autonomy and thus French-Canadian interests. In the wake of the conscription plebiscite (spring 1942), Chaloult broke definitively with the Liberals and helped to found the nationalist movement Bloc populaire canadien in the fall of 1942, a party with clear repugnance for Quebec's two established principal political parties and for the "economic dictatorship" that relegated French Canadians to a condition of poverty and submission.[10] Chaloult pursued his criticism of Godbout's economic policies in the months following the Legislative session of 1943. "It is the trusts that keep us in poverty. At all cost, we must liberate ourselves ... by nationalization or the exploitation by the State of a part of our natural resources," stated Chaloult in September. "The Union nationale and the Liberal Party solemnly committed themselves, in writing ... to nationalize and to combat the odious trust. They have done nothing about this. They will do nothing about this."[11]

Duplessis, too, kept up his assault on Godbout's laxity toward Aluminum Company through the summer and into the autumn of 1943. At Saint-Jean, before an audience of some 10,000 supporters, Duplessis harassed the Liberals for ceding Quebec's autonomy to a foreign multinational corporation as well as to the national government in Ottawa. "The Union Nationale has always refused this company ... what Mr Godbout has granted it," Duplessis charged. "The result is that the Americans and the Aluminum trust will easily remain the proprietors of our natural resources ... We must ask ourselves if we are in the province of Quebec or rather in the province of Aluminum."[12] At towns along the Saguenay in early October, Duplessis elaborated on these attacks. "The Union Nationale wants to treat capital with justice," he stated in Chicoutimi before 2,000 spectators. "But we do not want foreign money to direct the government ... The Aluminum Company mocked the rights of our province and this with the connivance of Ottawa and Godbout ... I said 'no' to Mr Powell ... But they have obtained Manouan, Passe-Dangereuse and Shipshaw for a mouthful of bread. They began their works before plans were approved and drowned thousands of acres of land including a large quantity of wood at a time when wood for heating is rare ... And while Godbout gave away our national heritage, Mr King exempted the Aluminum Company from $158 million in taxes ... Do we live in the province of Aluminum or in the province of Quebec?"[13]

In reading remarks such as Duplessis's and Chaloult's, one has to wonder how the wartime flak absorbed by Godbout over his treatment of Aluminum Company of Canada may have influenced his government's decision to create Hydro-Québec. The timing is certainly suggestive.

The Liberals' party platform of 1939 included an intention to nationalize the Montreal-based utility, Montreal Light, Heat and Power Company (MLHP). The following year, Godbout gave additional powers for public oversight of electrical utilities to the province's Electricity Commission, the now-named Public Service Board, which began a study to determine the legitimacy of the electrical rates charged by MLHP. Godbout's minister of lands and forests, Côté, requested a report from Hydraulic Service chief Latreille as to the possibility and financial advisability of expropriation. In May 1941, presumably with the information in hand, the government passed a bill authorizing the minister to acquire the power-producing subsidiary of MLHP, Beauharnois Light, Heat and Power, by expropriation if necessary.[14] However, no concrete steps were taken to carry out expropriation until the autumn of 1943 – some six months after Duplessis's most vehement and lengthy criticism of Godbout and Hamel in the Assembly, and within days of his scolding along the Saguenay. At that point in time, Quebec's Public Service Board ordered MLHP to publicly defend its rates and profits,[15] and Godbout dropped a political 'bomb' by announcing his intention to submit a bill in the upcoming legislative session to expropriate the firm. "We have decided," announced the agronomist-premier, employing a metaphor of the farm, "to take the bull by the horns."[16]

What explains this two-and-a-half-year gap? Why, having announced its intention to expropriate a portion of the Montreal utility in the spring of 1941, did the government fail to move on this decision until the autumn of 1943? Through 1942 and early 1943, the Godbout administration answered this question by explaining that expropriation was not advisable during wartime. Godbout may well have been following Ottawa's advice in this matter. Symington felt strongly that the war was no time "to settle fundamental questions" of private versus public ownership, especially as expropriation would threaten the pace of war production.[17] Reputedly, such views moved Quebec to delay nationalization. In May and September 1942, the *Financial Post* reported, "Ottawa has asked Quebec to put off any action [as the] Dominion government fears such a move at this time would ... be harmful to the war effort."[18] It is also possible that the Godbout administration was indecisive as to the form nationalization should take. As of 1941, the government's stated intention was to expropriate Beauharnois; in 1942, Godbout confirmed rumours that he intended to nationalize all the electric producers of the province, which may explain Alcan's anxiety in this regard; and at the outset of the legislative session of 1943, it was intimated that the unexploited powers along the Ottawa River would serve as the power base for a new provincial Hydro organization.[19] But when nationalization was actually announced and undertaken in 1943–44, the project was again readjusted: to expropriate MLHP. A final possibility for the delay is that the Public Service Board took considerable time to gather and prepare its case. This was the Godbout administration's explanation in the fall of 1943: that its announcement followed a three-year survey by the Board to review the account books and evaluate the assets of MLHP in order to determine the legitimacy of the rates charged. And that only with such information firmly in hand could the Godbout government justify taking the radical step of expropriation.[20] Yet Pigeon could inform Hamel of the investigation's results as early as February 1943, noting MLHP's dubious capitalization and "elevated profits."[21] Furthermore, as the government itself had earlier excused the delay in terms of the war and/or the Quebec-Ontario border agreement regarding power development, and certainly not in terms of the work of the Public Service Board, we should use caution in accepting the latter explanation at face value.

Much of the Godbout Papers material was damaged by water and, in any case, more politically sensitive material may well have been kept out of the public record. Barring more specific evidence, we should simply note that the Godbout administration took periodic criticism over its concessions of both Lac Manouan and Passe Dangereuse during the 1941–43 period and, as we will see, Aluminum Company's requests for additional concessions beyond the Saguenay watershed stirred deep resentment among the economic nationalists within the Godbout administration. By mid-war, due in no small part to Aluminum's expansion at the Saguenay, the control of hydraulic resources by English-speaking capitalists had again become a "red hot political question" in Quebec,[22] as it had been during the Depression era.

The fact that parliamentarians and the press, alike, had trouble distinguishing the Passe Dangereuse project on the Peribonka from the concurrent and gargantuan Shipshaw project on the Saguenay (whose rights had been sold outright at the turn of the century), only gave further credence to Duplessis's bloated and generalized claims of government "treason."[23] That 'Alcan' was still viewed as inseparable from American colossus 'Alcoa' (which had been under investigation by the

US government for monopoly control of the aluminum industry since the early 1920s[24]) also increased Quebecers' suspicion of the Canadian-American firm: whatever Alcan executives' claims of separate stock ownership and distinct corporate management, the company was perceived as but "the Canadian subsidiary of the American monopoly." The Arvida strike in the summer of 1941 cemented the impression among Quebec nationalists of an abusive "American monopoly" operating on Quebec soil.[25] The negative press coverage continued in 1943 as CCF leader Coldwell levelled charges in the House of Commons and beyond of Alcan's monopolization of the Saguenay's waterpowers and called for Shipshaw's expropriation.[26] By mid-war, there was some irony in the fact that Alcan was doing yeoman's work to win the war for the Allies – the Canadian multinational producing some 30 per cent of the ingot required by the United Nations – yet, in its home province of Quebec, the company was suspected of manipulating both provincial and federal governments, monopolizing strategic Canadian waterpowers on behalf of an American 'trust,' and engaging in wartime profiteering. And it was to this much-maligned company that Godbout had made the most significant resource concessions of his administrative tenure.

In this political context, it would certainly have been useful for Godbout to establish a more proactive and progressive record in the management of the electricity industry, and to do so in advance of a soon-approaching election.[27] The business-friendly English-language press had no doubt that the premier's action was a political one. "Godbout is setting the stage for an election," charged the *Financial Post*, his plan being to "spike the guns of his political opponents who have agitated for nationalization of the company, largest privately owned utility in the Dominion." *The Gazette* concurred: "The announced plans bear all the earmarks of having been hastily spawned in an atmosphere of political hysteria ... They have been trumpeted as a catch-cry to bolster the sagging support for the government in the election expected early next year." Furthermore, "desperate ... for a political football to kick around in the next election," *The Gazette* held that Godbout was willing to sacrifice "justice and the protection of the law."[28] Godbout's French-speaking enemies suspected a political motive as well. Duplessis, of course, was quick to characterize Godbout's nationalization plans as but "une mesure électorale" calculated "to catch votes."[29] Robert Rumilly, historian-polemicist and admirer of Duplessis, intimated that Hydro-Québec's creation was the inspiration of T.D. Bouchard, who saw the measure as a means to offset criticism of Godbout's Aluminum policies. Wrote Rumilly (albeit without citing evidence): Bouchard envisioned "a great coup for the next session – that must precede the elections."[30] And Godbout's announcement did generate real excitement among Quebec nationalists. "All bombs aren't disagreeable to receive," *L'Action catholique's* Eugene L'Heureux editorialized wittily. "The one that exploded in the province of Quebec Friday evening was evidently welcomed with joy by the majority of our citizens … It is [also] the most effective blow ever delivered to the electrical trust."[31]

This chronology suggests that the war itself could have been an important trigger for the nationalization of Quebec's electricity industry. And one need not be a member of the Conservative/Union Nationale camp to imagine this cause-and-effect relationship. Certainly the root causes of nationalization are found in earlier French-Canadian nationalist demands, from the turn of the century, to wrest control of the development of crucial hydraulic resources from an English-speaking economic elite. From such nationalists' calls grew a political movement in the 1930s against the power 'trust' and the appointment in 1934 of a board of public inquiry, the Quebec Electricity Commission, a predecessor of the Public Service Board. A decade later, in the spring of 1944, the government of Godbout fashioned the Commission hydroélectrique de Quebec (Hydro-Québec) to expropriate and/or develop waterpowers and transmission systems under the supervision of the state and in the interests of the province.[32] But what was it that tripped this final and conclusive event in the formation of Hydro-Québec? It is not sufficient to note the Godbout regime's penchant for progressive policy.[33] For one thing, Godbout appears far more progressive in hindsight, and when judged by the legislation his government left behind, than in the view of his contemporaries. For another, Godbout was too intelligent and receptive a politician to be guided by ideology alone in a decision of such magnitude. Moreover, the specific context of Hydro-Québec's birth merits attention and examination by historians without simply accepting, as though inevitable, the unfolding of this historic event. As Godbout himself stated, the timing of the decision may partly be explained by the Public Service Board's report on the high electrical rates and dubious accounting practices of MLHP.[34] When did the Board actually have results available, we might wonder? Did the Board require three years of investigation to conclude that MLHP was charging excessively high rates?

Or did the sensitive politics of hydroelectricity, as driven by the crisis of wartime, also play a significant role in Hydro-Québec's creation?

Historian Matthew Evenden, in his review of the wartime mobilization of Quebec's rivers, suggests that the establishment of a federal power controller in 1940, supreme in matters of electrical production and distribution, "weakened provincial jurisdiction" over the regulation of waterpowers.[35] This may well have been so. In the case of Passe Dangereuse, while not actually weakening provincial decision-making, the war certainly *accelerated* it in the matter of hydraulic power, and in a manner that produced considerable political backlash in the province. In this way the Second World War, and the government's active response to that emergency in promoting dam building by a private firm, helped to create a political climate that was briefly ripe for public power. Godbout may have put off nationalization due to wartime, while also awaiting a formal report by the Public Service Board on MLHP, but holding to that position became untenable by the summer and autumn of 1943. The emerging political pressure for bold action on the hydro front – which might help counter broader Opposition and nationalist charges that Godbout served as Ottawa's puppet (Ottawa's "Charlie McCarthy," as Duplessis savagely put it) – pressed the administration to abandon its wait-out-the-war strategy and take the radical step of expropriation. So, while wartime at first delayed nationalization, the war eventually forced the issue as politically expedient and even urgent. By such subtle and indirect means, the Peribonka concessions likely helped to catalyze the Godbout administration's decision of late 1943 to go forward definitively with the establishment of Hydro-Québec.

6

Bersimis and Beyond

The history of hydroelectricity across the twentieth century can be written in terms of increasing geographic scales of power exploitation: from the development of isolated falls, to the exploitation of a river's full 'power reach,' to the use of multiple watersheds via river diversion. This growing scale of development was abetted and encouraged by technological advancements (in dam construction, turbine design, power transmission), by urbanization and electrification (including growing consumer and industrial demand for electricity), and, of course, by available venture capital.

The exploitation of Quebec's Saguenay basin provides a case in point. At the turn of the century, reflecting a formative period of hydroelectric technology in the 1890s, Canadian entrepreneurs such as Quebec City timber merchant Benjamin Scott and Ontario inventor Thomas Willson envisioned using the head of particular waterfalls or limited sections of the Saguenay for a singular commercial purpose. These relatively small-scale plans were foiled for want of adequate venture capital or power customers in a remote and rural North American outland. By the era of the First World War, with arrival on the scene of American industrialist J.B. Duke, the Saguenay experienced a shift in developmental scale. It was Duke's hydraulic engineer William States Lee who embraced the use of Lac St-Jean as a natural storage 'pond' to regulate the flow of the Saguenay below. It was also Lee who conceived of amassing the Saguenay's various falls and rapids in two giant dam-built steps (now Isle Maligne and Shipshaw) to exploit the river's entire 300-foot head and produce the maximum and constant quantity of electricity required for world-class electrochemical manufacture. The Carolina engineer also studied the Peribonka River's exploitation as a means of additional flow regulation of the entire basin. The 1920s saw the Duke organization's completion of the upper Saguenay (Isle Maligne) scheme with use of the enormous storage capacity of Lac St-Jean.[1] By the 1940s, Aluminum Company was engaged in building the Shipshaw plant along the lower Saguenay, and, high in the Peribonka valley, impounding Lac Manouan and Passe Dangereuse.

Then, midway through the Second World War, Aluminum Company's executives and engineers moved to burst the bounds of the Saguenay basin by calling for the impoundment and diversion of neighbouring rivers.

Their motive, as before, was increased firm power production. Throughout the first half of the war, Canada's power controller, working in co-operation with Quebec's Public Service Board (PSB), had directed the mobilization of power resources throughout Quebec to serve aluminum production. Getting Shipshaw built was certainly their most significant contribution toward adding new power capacity. Symington and the PSB also orchestrated the construction of transmission lines in order to convey power between Quebec's major hydroelectric systems of the St Maurice, St Lawrence, and Saguenay valleys. The impressive result, as Shawinigan Water and Power Company's electrical engineer described it, was to increase the province's "combined firm power capacity far beyond the sum of the capacities of the individual systems if operated independently" and the creation of "one of the major power systems on the continent, having resources of nearly 4,000,000 horsepower" developed or under construction. Symington and the PSB also oversaw the diversion of power from newsprint manufacturing, the upgrading of existing hydroelectric plants by private firms, and the domestic conservation of energy.[2]

Despite these many efforts, soaring Allied demand for ingot threatened to outpace power production in Quebec and the entire northeast region, causing considerable worry to power experts on both sides of the border. As of early 1942, the PSB's McCammon felt that Quebec's power situation was "satisfactory for the present if good water conditions prevailed, but not otherwise." Symington was less sanguine, pointing out that the US government was again in conversation with Aluminum Company of Canada "about doubling present plans for producing aluminum," which would involve twelve generators at Shipshaw, rather than six, and "which would necessitate expensive new storage."[3] In the spring of 1942, Symington understood that "power would be a limiting factor" in the production of aluminum. With Shipshaw now rising ahead of schedule, Powell and Dubose apparently did not share the power controller's anxiety about major power shortages in the coming winter. Powell stated that limited bauxite supplies constituted "a much more serious danger" than anything else. But Aluminum executives did acknowledge that, in order to fulfill obligations to the United States, the firm would likely require at least 50,000 horsepower on top of the electricity generated at the company's own facilities and at other sources in the province.[4] "There will be a period during this winter [1942–43]," Dubose anticipated, "when [the Saguenay's aluminum] works will require more electricity than the Saguenay can produce – yes, more electricity than the Province can produce."[5] Even with Shipshaw under construction, with the province's electrical systems interconnected, and with new reservoirs forming, there loomed the possibility of a power shortage for the coming winter.[6]

In an effort to make up this shortfall, Aluminum sought to divert additional water into the Saguenay basin. The Bersimis River lies immediately to the east of the Manouan, flowing not into the Saguenay basin but south into the St Lawrence. And as we have seen, the Bersimis, like the Peribonka and Manouan rivers to the west, was still thoroughly occupied and utilized in this era by Native Peoples, whose summer home and trading post was situated at the Bersimis Reserve on the St Lawrence River.[7]

In May 1942, once the winter snow had melted and the rivers ran free, a team of Aluminum Company engineers led by Fred Lawton conducted aerial surveys of this region, followed immediately by a ground survey to locate dam sites. They identified the upper Bersimis as the most feasible and cost-effective place to gather additional water volume for the Saguenay's turbines. By month's end, Aluminum's executives in Montreal had approved the project.[8] So in just three weeks' time, Aluminum had explored, roughed out the design of, and approved funding for a method of upgrading the power system of the Saguenay, with the intention of capturing, if possible, the autumn rains of 1942. Dubose then begged minister of lands and forests Côté and Hydraulic Service chief Latreille to concede the necessary rights.

In these initial requests, Aluminum sought to flood but 15 square miles and capture some 600 square miles of drainage territory, representing 8 per cent of the Bersimis watershed. The water would be dammed and held in "a small storage reservoir" (of some 7.6 billion cubic feet) at Lac des Prairies and Lac Manouanis by means of stone-filled cribs and earthworks, and then diverted by a short westward-leading canal into the Manouan River and thence to the Peribonka, at least for the duration of the war. Dubose explained that at some point in the distant past, the waters of Lac des Prairies had fed the Manouan

naturally, which should make the government's "immediate acquies-
cence easier" since Aluminum would be returning flow conditions to
their "natural situation." As most of the territory lacked merchantable
timber (thus it was "water and waste lands"), and there were no Crown
timber limits leased in the vicinity, Dubose assured the state that the
project "will cause no damage to third parties." As no roads accessed this
region, construction would be supplied by airplane, as at Lac Manouan.
The dams and canal would cost the company an estimated $300,000, or
one-tenth of the Passe Dangereuse project cost, and might yield 15,000
horsepower during the winter of 1942–43 and 20,000 horsepower in
winters thereafter. Aluminum offered to pay $2.50 per cord for any
merchantable timber submerged by the new reservoir (about five square
miles), as well as $1.00 per horsepower on the resulting energy gener-
ated at the Saguenay power stations downstream. Dubose made clear
that it was the "pressing need" for power this winter driven by the "war's
emergency" that prompted this new project. He attempted to buttress
this argument by explaining that, "Mr Symington … states unreserv-
edly that we need the power." Having butted heads for many months
with Latreille over Passe Dangereuse, Dubose understood that there
might well be "difficulties in the way of granting the requested rights."
In submitting the necessary maps and graphs to permit the government
to make sense of the project, Dubose wrote, "I hope that you and Mr
Côté will go to the limit in aiding us to put this plan into effect," and he
prayed the government's "immediate consideration."[9]

The government would not be rushed. Latreille considered it
"irrelevant and immaterial" that the Bersimis may have once fed the
Manouan, since "even if proven, the existing situation was not brought
about by the work of man."[10] The Hydraulic Service chief considered
the diversion project analogous to that of the upper Mégiscane River
in western Quebec, which had recently been diverted by the Quebec
Streams Commission to add water volume to the Gouin Reservoir and
thence to the power plants along the St-Maurice River. The Mégiscane
project had required a special law passed by the full Assembly. Latreille
and Côté brought the Quebec Streams Commission into this discussion[11]
and also sought the advice of Louis-Philippe Pigeon in this regard. The
legal clerk of the Legislative Assembly confirmed that the grant would
require legislation because it represented a concession of waterpowers
in excess of 300 horsepower. Latreille (writing in the name of the minister) then informed Dubose, telling Aluminum's engineer that since
the Assembly's session had come to an end, the Department of Lands

and Forests had "no authority to grant Aluminum Company of Canada, Limited, the right to proceed with the said works."[12] Aluminum's attorneys, Geoffrion-Prud'homme, vehemently disagreed, arguing that the lieutenant-governor-in-council "had the necessary power to authorize this diversion" under current Quebec statute, not only because current law allowed for the storage or "deviation" of waters without special legislation but also because the Quebec Streams Commission (a public entity) was covered by a different body of law than were corporations. At the very least, given that "there is great urgency in connection with war work," Aluminum's attorneys and officers pressed the Godbout administration to authorize immediate construction (as it had done for Passe Dangereuse), pending final confirmation and approval of the project by the legislature at its next session.[13]

In an attempt to settle the matter, Aluminum's executives also urged the government to bundle the various issues "of a controversial nature" and resolve them all at once in a grand bargain. These matters included Passe Dangereuse's timber compensation and power tax, the diversion of water from the Bersimis River, and, interestingly, any future plans by the Quebec government to nationalize electricity. As V.E. Bird's memorandum put it, Aluminum sought the government's "assurance that ... all properties belonging to Aluminum [and its subsidiaries] ... will be exempted from expropriation when, as and if the Government becomes vested with the power to institute expropriation proceedings against public utilities."[14] These words suggest that Aluminum's executives were aware of and anxious about the political winds now developing in the province. Following this memorandum, Powell asked the premier for an interview.[15] There is no evidence that such a meeting took place, and no one in the government formally replied to Powell. The government did respond regarding Aluminum's request to waive parliamentary legislation for a diversion scheme: the Cabinet concurred with Latreille and Pigeon that a law must be passed.[16]

By the end of June, Aluminum had abandoned its plans to divert a portion of the Bersimis River for a much more ambitious project.[17] The newly proposed reservoir in Lac Pipmuacan (at 330 billion cubic feet) would exceed Passe Dangereuse in volume and would cost $6 million to $12 million to create. It would draw its waters from the impoundment of the upper aux Outardes River, via Lac Pletipi, capturing 22 per cent of this watershed; from the Bersimis River, at Lac Pipmuacan, representing some 70 per cent of this basin; and "possibly [from] the Manikuagan [Manicouagan River] as well." Once the lake was raised roughly one

hundred feet, its waters would be diverted by a batch of control dams
and canals into either the Manouan/Peribonka drainage to the west,
or into the Shipshaw River to the south, which enters the Saguenay at
the site of the hydroelectric plant that bears its name. Dubose touted
multiple advantages to this multi-watershed scheme, not the least of
which was that it "would increase the firm continuous horsepower on
the Saguenay by at least 150,000 hp." Even at a cost of $10 million,
"the power would be less expensive than at any other development that
could be made in Canada so far as I know." Since reservoirs, rather than
power plants, were the intention, Aluminum could avoid lengthy wartime waits for electrical and hydraulic equipment. Finally, the plan could
be executed quickly, perhaps in one year's time, with the diverted waters
of the aux Outardes and Manicouagan accelerating the filling of Lac
Pipmuacan.[18] Aluminum pursued its aerial and ground surveys through
the summer. In mid-August, the company confirmed its desire to go
forward with the Bersimis/Outardes project and asked the government
for immediate authorization to commence work pending Legislative
approval the following spring.[19]

Latreille was unmoved by Dubose's arguments and, in fact, held
"grave objections" to the project. For Latreille, the "immense storage
reservoir" at Lac Pipmuacan would "upset the natural conditions of a
large extent of the north shore of the St Lawrence." These conditions
included timber, hundreds of thousands of acres of which would be
submerged before it could be cut. One important timber company,
Brown Corporation, already held timber limits in the Bersimis valley
and was negotiating additional leases, so their operations might well
be "gravely compromised" by a hydro concession to Aluminum. The
province's hydraulic wealth, too, would be damaged by a project that
intended to change the "natural flow conditions" of several important
rivers. The fact that Quebec North Shore Paper Company operated a
power plant of 77,300-horsepower capacity near the mouth of the aux
Outardes River was immediately problematic. Latreille was also aware
that the Hudson's Bay Company operated a trade in furs in the northern
reaches of the Saguenay and Côte-Nord rivers, although the fur trade
and its participants did not figure with any significance in his concerns.

Latreille's overarching concern was that the wartime power needs of
the aluminum industry not damage or prejudice the general provincial
economy, as long as power could be obtained by some other means.
Rapide Trenche on the St-Maurice River, for example, was currently
under lease to the Shawinigan Water and Power Company, but was not

yet developed. Latreille calculated that the costs of building a hydroelectric plant there, the resulting energy yield, and the time to bring
this project to fruition were all comparable to the Bersimis/Outardes
storage scheme. But whereas the Bersimis/Outardes diversion would
likely only serve Aluminum for the duration of the war, Rapide Trenche
would serve the province for the long term, providing power for industry to the St-Maurice valley or via transmission to Lac St-Jean – and no
law needed passing, since the Shawinigan company already held the
concession. Aluminum Company might also complete the excavation
of the gorge of the Grand Discharge at the mouth of Lac St-Jean, which
would give the Company access to additional storage in the lake and
raise power production by 75,000 horsepower. Latreille saw "danger"
in allowing a single corporation to command such "a vast concentration of electrical energy." And he held serious doubts that Aluminum
Company intended to promote economic development for any greater
good. "Aluminum has shown … just how little it cares about economic
development," Latreille wrote the premier. "The new development at
Shipshaw will render useless the development at Chute à Caron, and by
consequence this development for which we are currently obtaining all
sorts of [wartime] priorities so that Aluminum can use 210 feet of water
rather than 150 at Chute à Caron, is an installation whose productivity
is hardly in accord with its expense."[20]

Prosecuting the war may have been important, but for Latreille the
long-term economic development of Quebec was of higher priority.
From the war's start, he balked over further power exports to Ontario,
lest the war serve to grow the economy of the rival province at the
expense of Quebec. Therefore, new power resources developed in
hydro-rich Quebec must be exploited with an eye "to the benefit of
our province." The Hydro Service chief had also bristled at the suggestions made by the American delegates at the Canadian-American
power summit meeting of February 1942, including the suggestion that
Quebec sharply reduce delivery of power to pulp and paper mills to
conserve power for the aluminum industry. Latreille felt strongly that
any measures taken in the service of the war must also account for the
"grave repercussions" of limiting the industrial expansion of Quebec
vis-à-vis neighbouring Ontario.[21]

Economic nationalism also imbued the response of Pigeon.
Godbout's brilliant legal adviser had been a forceful advocate for reform
of the hydro sector and may properly be considered the architect of
Hydro-Québec. It was Pigeon who sketched out the design of the new

Hydro-Electric Commission of Quebec[22] and then drafted the law creating Hydro-Québec. From the outset of Godbout's administration, Pigeon had pushed the premier to assert genuine control over the electric utilities and to distinguish himself from his permissive Liberal predecessor Taschereau, a vigorous promotor of private enterprise.[23] In Pigeon's estimation, Taschereau's merely moderate reform in the hydro sector (the 1935 creation of a regulatory body, the Electricity Commission, to more closely supervise private industry) was judged by the electorate as political "camouflage" rather than "a sincere effort"; and the voters' instinctive distrust of such Liberal lipservice to progressive governance, contributed significantly to the defection of the youthful wing of the party (in the formation of the Action libérale nationale) and to the Liberals' catastrophic loss of seats in the election of October 1935 and a much-reduced majority in the Assembly. Godbout, by contrast, had won power in 1939 because be bore the electorate's confidence that he was a "new man" ("homme nouveau") untainted by any sordid political past. Therefore, if Godbout hoped to retain the confidence of Quebec voters, he must bring real and substantive change to the electricity industry. "It is of supreme importance," Pigeon had urged Godbout in 1940, regarding hydroelectric policy, "that you do not give the impression that you are re-establishing the regime of pre-1936."[24]

In the current matter of the North Shore diversions, Pigeon had been brought into the discussion to provide legal advice on the necessity of legislation. Pigeon took it upon himself to contribute a full-fledged indictment of the Godbout administration's policy regarding Aluminum Company of Canada and the power industry generally. Aluminum Company has obtained "favourable treatment without precedent" in the concession of the two storage reservoirs, Pigeon told the premier. In the case of Lac Manouan, the company enjoyed a modest rent in lieu of a more remunerative power tax; and in the case of Passe Dangereuse, the government was showing itself to be permissively flexible regarding the length of the lease, the compensation to timber interests, and the manner of imposing a power tax. The previous Liberal administration of Taschereau had already favoured the company with generous tax terms on Lac St-Jean, Pigeon argued. As a consequence, Aluminum Company paid what amounted to 16 cents per horsepower/produced for the energy it generated at the Saguenay. In contrast, Shawinigan Water and Power paid 78 cents per horsepower/produced from the storage it enjoyed in the Gouin Reservoir, even while helping to attract a host of electrochemical and other power users to the St-Maurice region.

Weighing heavily on Pigeon's mind, we should note, was the fact that
Aluminum Company had begun construction that summer of a large,
new aluminum foundry on a forty-eight-acre site just outside Toronto.
For the company's executives, it was quite logical to replace an aging
urban facility (the Toronto Works, 1913) with an increasingly mecha-
nized suburban one.[25] For Pigeon, it represented betrayal. Aluminum's
competitive advantage, after all, derived from cheap waterpower in
Quebec, which in turn derived from Liberal generosity to the firm. "In
return for these favours [the Aluminum Company] installs factories in
Ontario."

Pigeon also reflected on the larger costs of the government's generosity to the power industry. "If one day we decide to expropriate the
electric power producers, we will have to pay as indemnity an amount
that includes all the millions that we've abandoned to companies by such
favourable treatment ... It seems to me that it is high time to change
the policy and to cease all favourable treatment to those that exploit
our natural resources."[26] It wasn't only private corporations that Pigeon
and Latreille distrusted in matters of resource development, but also
the governments of Ontario and Canada. Both men jealously guarded
the province's hydroelectric resources. In Quebec's exports of power to
Ontario, in the sharing of Ottawa River hydro sites with Ontario (agreed
to in 1943), in Ottawa's construction of war plants in Ontario to utilize
Quebec energy, and in the future development of the St Lawrence, they
saw potential threats to Quebec's economic future from rival provincial
and national regimes.[27]

Minister Côté asked the other members of Quebec's hydro-bureau-
cracy to weigh in on the diversion scheme; once they had studied
the proposal, all were unified in their opposition.[28] McCammon and
Normandin of the Public Service Board explained that the Quebec
state as yet lacked adequate hydrographic information on the drainage basins in question; they recommended lengthy and detailed
studies before negotiations could begin in earnest. The government
should "immediately show its opposition" to the enormous project, and
Aluminum should be told to look elsewhere for energy. Lefebvre of
the Streams Commission shared this view. In fact, he and Latreille had
a telling phone conversation in early September, the record of which
is fortunately preserved in a memorandum. In this conversation, the
two engineers referred to Aluminum Company of Canada as "Alcoa"
and when Lefebvre expressed a willingness to grant "Alcoa" a certain
or partial diversion of the Bersimis, Latreille told him that "once the

principle was adopted, Alcoa will succeed by stages to get all that it wants," as it had done with both Lac Manouan and Passe Dangereuse.[29] Aluminum's aggressive reach into the basins beyond the Saguenay was angering, perhaps even radicalizing, several members of the Godbout administration; these men evidently equated Alcan's expansionist plans with a larger American economic invasion.

The group of hydro engineers – Latreille, Normandin, McCammon, and Lefebvre – met with Pigeon in mid-September to orchestrate their opposition to the Bersimis/Outardes diversion.[30] Latreille wondered aloud with minister Côté whether Aluminum's wartime needs for power were entirely genuine, or perhaps "pretended," and he made every effort to gather the necessary data and/or provincial power forecasts from the Public Service Board in order to make up his mind.[31] For his part, the Quebec Streams Commission's Lefebvre considered the supposed power shortage non-existent.[32] The Liberal press aired such suspicions publicly. *Le Soleil* sounded the "alarm" that Aluminum and the power trusts were conducting a "campaign" to convince the people and the government of power shortages. Aluminum was using the excuse of the war emergency to grab additional electrical resources in the province. Aluminum's expansion, *Le Soleil* argued, would in turn exacerbate "the problem regarding the economic liberation" of French Canadians. "While the people consent to these concessions in order to aid the war effort and liberate the world, the magnates of international enterprise will be placing them in new chains."[33] The story was picked up along the Saguenay as well. Chicoutimi's *Le Progrès du Saguenay* wondered "what formidable power" Alcan would command due to the wartime concessions of Lac Manouan and Passe Dangereuse, power that the company "will retain after the war."[34]

Godbout's ministers Oscar Drouin and Télesphore-Damien Bouchard, long-time protagonists of hydroelectricity's nationalization, echoed these concerns. As Cabinet members, they were fully informed as to Aluminum's requests for water rights beyond the Saguenay basin. They, too, feared that the war was providing Quebec's regional power monopolies with the opportunity to expand and extend their territorial control and influence. Bouchard and Drouin's response to these events was to publicly reiterate their desire to see hydroelectricity nationalized in Quebec.[35] Drouin argued that Quebec's natural resources were the source for French Canadians' "richesse nationale." But as long as the present economic system endured, never would the race obtain the benefits of this wealth. Radical change was therefore necessary. "It's not a matter of confiscation or destruction. This would be repugnant to our people. But it behooves the people, by legal and honest means, to find a way to create national wealth of which the fruits will go to all and not only to a few individuals." We must have a "politique pro-cana-dienne-française," asserted the minister of commerce and industry, and there cannot be any other policy.[36] "Change occurs not only by evolution but by revolution," agreed Bouchard. "We cannot wait for important change such as the nationalization of electricity ... The time has come for the province of Quebec to have its Hydro Electric that will provide us with lower rates for both domestic and industrial power consumption."[37] Bouchard had served as the mayor of St-Hyacinthe for a quarter of a century, and under his leadership, the town had municipalized electricity. Now, with Bouchard's guidance, the town's Chamber of Commerce passed a resolution recommending that the provincial government follow St-Hyacinthe's lead by creating a provincial hydro organization to serve the Montreal region.[38] The province-wide Fédération des chambres de commerce followed suit. Meeting in Sherbrooke, the Fédération endorsed Bouchard's call for a "Hydro-Électrique Provinciale."[39]

Such hostility toward private enterprise, and Aluminum Company in particular, soon crystallized in a decision on the Bersimis/Outardes proposal. In early October 1942, Quebec's hydraulic experts met in Montreal with the officers of Aluminum Company. The Hydraulic Service chief had long signalled to Aluminum the government's "objections to so considerable a modification of the natural conditions of [the Northshore] region."[40] At this October meeting, the government definitively refused the Pipmaukin diversion, but did not categorically refuse a smaller-scale diversion project, say, of a portion of the aux Outardes River alone via Lac Pletipi, representing some 22 per cent of the aux Outardes watershed. For some months, Aluminum still held out hope for the larger project and evidently exerted "pressure" in Ottawa for intervention in Quebec.[41]

Meanwhile, the Hydraulic Service began to investigate the more discrete Lac Pletipi diversion scheme – and quickly encountered strong opposition from the paper industry.[42] The Quebec North Shore Paper Company was founded by Robert McCormick's *Chicago Tribune* to augment the newspaper's supply of newsprint. In 1938, the firm opened its newsprint mill (the only one built in Quebec during the Great Depression) near the mouths of the aux Outardes and Manicouagan rivers, drawing electricity from the company's power dam at aux Outardes's lowest falls. Thus was born the resource town of Baie-Comeau.[43] Quebec

North Shore's president and general manager Arthur Schmon put it forcefully to lands and forests minister Hamel that "diversion constitutes the most radical infringement of the basic law of flowing waters." Quebec North Shore had leased provincial water rights in good faith during the 1920s, he pointed out. He argued that diverting nearly one-quarter of the aux Outardes River at Lac Pletipi would "completely change the character of the river," especially since the lake constituted the most important natural storage facility in the watershed. Thus the proposed diversion "would make it impossible to supply continuous and dependable motive power to the Baie-Comeau mill." It would also lead to the "destruction and confiscation of an investment of 20 million dollars," comprising the power and manufacturing facilities that made up this enterprise. Nor was this private, US-backed industry alone at risk. Also in jeopardy, Schmon pointed out, were the factory's 500 workers and a town of 2,500 people, along with its schools, homes, churches, and markets for farmers and merchants. Schmon urged that the "Government should avoid the infliction of such hardships as long as the applicants for such a diversion have other possible means to provide for the threatened shortages which may or may not materialize." Schmon himself made several suggestions as to where Aluminum could find more power, the "complete and final solution" being the diversion of the voluminous rivers of James and Hudson's bays into the St Lawrence drainage (the Rupert River's giant Lake Mistassini, for example, might be diverted via Lake Chibougamau to the Ashuapmushuan River to Lac St-Jean), thus making landscape changes to a more remote region where industries were not yet established, and (he might have added) only 'Indians' made their home.[44]

Schmon's vigorous objections found a receptive audience in the Department of Lands and Forests. War or no war, Latreille and Hamel would not dare to damage an entrenched Quebec industry along with the French-Canadian community it supported. By the end of the 1942 calendar year, the department had refused Aluminum Company any and all diversions of water into the Saguenay basin from the Bersimis, aux Outardes, or Manicouagan rivers.[45] Denied exploitation of the Saguenay's neighbouring rivers, Aluminum Company began casting about elsewhere for large sources of hydroelectricity. By early 1943, Dubose was expressing interest to the government of British Columbia in a large block of power. This interest would bear fruit in the postwar period in Aluminum's gargantuan diversion, power, and smelting project known as Kitimat-Kemano.[46]

7

Conclusion

Three decisions about the course of rivers – the Manouan, the Peribonka, and the Bersimis – provide us with something of a measure and range of what the Quebec government found acceptable to concede in wartime. And those decisions seem to have depended largely on the scale and nature of capital's demands. Lac Manouan's remoteness meant there were no competing forestry interests involved, and it drove up the cost and increased the difficulty of dam-building. The pre-existence of a chain of lakes in the region meant relatively minor rearrangement of a natural order. All of this produced a relatively easy and generous concession by the state. Conceding the enormous Passe Dangereuse reservoir on the upper Peribonka River was also deemed necessary in wartime, and the project presented promising revenue possibilities in the form of a long-term lease and power tax. But the tax calculation, as well as the pre-existence of forestry interests in the region, served to slow the negotiation process and toughen the agreed-upon terms. The Bersimis diversion went beyond the bounds of what Quebec's waterpower experts were willing to tolerate. The warming of Quebec hydro nationalism by mid-war – in part an outgrowth of the rising hostility toward Aluminum Company – may also have played a role in the outright rejection of the North Shore demands.

The Peribonka concessions might shed light on the resource politics of two provincial regimes, but characterizing the decision-making styles of the Duplessis and Godbout eras defies simple generalization. Duplessis, long understood by historians as a "staunch champion of industrial capitalism,"[1] especially during his second administration of 1944–59, is not so easily typecast in the matter of the Peribonka. Not only did he correctly reject a bad and unnecessary deal with Alcan, but he did so having consulted, at least cursorily, the available waterpower expert/ civil servant Normandin before handing down judgment. Conversely,

Godbout, the agronomist-turned-progressive-reformer (advocate of women's suffrage and universal education, founder of Hydro-Québec), actively made concessions to Aluminum Company in the service of wartime's dire need, going so far as to waive provincial law to get the dams built and the water stored in rapid time. Duplessis and Godbout seem to have made practical, rather than ideological, decisions in the case of the Saguenay basin's hydroelectric development. Circumstances, more so than any political label as conservative or reformer, seem to have guided the policy actions of these two administrations.

That said, there is much to typecast in these two administrations with regard to the *style* of political negotiation. Duplessis's personal intervention in the matter of Lac Manouan nicely matches scholars' portrait of the "Little Dictator," contemptuous of political process that he himself did not command.[2] And what Alcan experienced in the late 1930s, it would again encounter in the post-war period as the firm sought the Duplessis government's authorization to develop additional hydroelectric power plants along the Peribonka River.[3] Similarly, Godbout's willingness to meaningfully involve his ministers and civil servants nicely illustrates historians' characterization of him as a politician who delegated rather than dictated the application of policy,[4] and did so even in the midst of the crisis of wartime.

Where Quebec hydro policy is concerned, the Peribonka concessions introduce the possibility that war – specifically the political crosswinds of Canadian-American wartime mobilization – served to catalyze nationalization in 1944. Our evidence is suggestive and circumstantial rather than conclusive, but it is nevertheless intriguing. In brief: the war accelerated and also softened decision-making under Godbout, who provided Alcan with a generous concession at Lac Manouan and then permitted the company to commence construction of Passe Dangereuse before obtaining provincial authorization. Godbout's lieutenants, including political heavyweights Bouchard, Drouin, Chaloult, and Pigeon, fiercely criticized such government-granted 'favours' and pressed the premier to make amends by acting forthrightly in the creation of provincial public power. Also critical were seasoned civil servants/hydraulic specialists Latreille and Lefebvre, who considered Alcan synonymous with the much-maligned US firm Alcoa. More potent still were the attacks by the Opposition. Inspired by the US-based Shipshaw scandal and its Canadian counterpart, Duplessis harassed Godbout throughout the summer of 1943 for coddling the Canadian-American aluminum giant. Duplessis also charged the government with gross hypocrisy for having promoted hydro's nationalization in the 1939 campaign. The Union Nationale chief, employing his characteristically caustic rhetoric, lashed out: "Do we live in the province of Aluminum or in the Province of Quebec?" In the immediate wake of Duplessis's charges, and contemplating an election, Godbout announced his intention to form the Hydro-Electric Commission of Québec (Hydro-Québec) to expropriate and operate the energy assets of Montreal Light, Heat and Power.

To be sure, the Alcan controversy of 1942–43 was likely not the only trigger for nationalization. Quebec concluded an interprovincial agreement with Ontario early in 1943 regarding the allocation of undeveloped hydraulic resources along the Ottawa River. The accord granted Quebec the rights to power sites near the city of Montreal, including the site at Carillon, which might eventually be developed by the new provincial utility. Perhaps more important, the Quebec-Ontario accord, while negotiated rapidly, smoothly, and fairly between March 1942 and January 1943, nevertheless, provided Duplessis with yet another opening for a hyperbolic, bombastic attack on Godbout. The Union Nationale chief accused the Liberal government of "one of its worst treasons by ceding to the province of Ontario a portion of our rich territory," etc.[5] The Public Service Board's study of Montreal Light, Heat and Power must also be factored into this chronology. During these same early months of 1943, Godbout would have had access to this provincial study of MLHP's corporate account books and thus information about the firm's assets and profits.

Moreover, Hydro-Québec was born of a broader historical phenomenon, one that spanned the Canadian-American border and was shaped by the Progressive Era, the Great Depression, and the Second World War. This was the Progressive Era's public power movement, whose advocates argued that electricity should be a not-for-profit 'service' that was owned and regulated by government for the benefit of its citizenry, rather than a 'commodity' bought and sold for corporate profit.[6] Already well advanced in Europe, the movement found traction in Ontario and Manitoba as early as 1906 and in Nova Scotia and New Brunswick by 1920.[7] Of particular import was the creation of the Ontario Hydro-Electric Power Commission (Ontario Hydro). Its creation was driven by a broad coalition of Ontario businessmen, civic leaders, union members, and social reformers who were concerned that private power exports to the United States (from the Canadian side of Niagara Falls) threatened to stifle the province's industrial/economic development. In turn, Ontario Hydro's successful, large-scale experiment in public ownership (by 1930, it was the world's largest hydroelectric system, serving 75 per cent of Ontario's electricity needs and commanding an international engineering reputation[8]) inspired progressives in the United States. These included New York's governor Franklin Roosevelt who signed the New York Power Authority into law in 1931 "to give back to the people the waterpower which is theirs."[9] As the Depression moved American society to the political left and FDR became president, the US public power movement grew in influence and extent: the New Deal era saw creation of the Tennessee Valley Authority (1933) and Bonneville Power Administration (1937) to construct hydroelectric power systems in the Tennessee and Columbia basins. Meanwhile, the Second World War prodded governments to assume increasing control over or direction of industrial production, including the production of electricity. Quebec was among several provinces (notably British Columbia and, to a lesser degree, Alberta)[10] in the throes of a wartime public power debate. In British Columbia, that debate culminated in 1945 when the provincial government created the British Columbia Power Commission, which would grow into the powerful BC Hydro in the early 1960s. Similarly, Quebec's first step in creating a provincial public utility in 1944 would blossom into full-scale nationalization in 1963.

Of course, the tone of Quebec's debate was unique in Canada, imbued as it was with the rhetoric and emotion of ethnic nationalism (as opposed to, simply, anti-monopoly populism or provincial nationalism). French-speaking Quebecers were pursuing a 'politique pro-canadienne-française' and ultimately 'economic liberation' from an English-speaking and mainly American corporate elite. However, the timing of Quebec's wartime nationalization was unremarkable, falling as it did squarely into an era of expanding public enterprise during the first half of the twentieth century.

Without more concrete evidence of Godbout's motivations, it is difficult to isolate these several streams of influence from one another or to assign them an order of importance. It may well be that the Peribonka's politics triggered Godbout's decision on nationalization in 1944. Just as likely, a more complicated confluence of catalysts and root causes moved Quebec's wartime Liberal government to establish Hydro-Québec.

Looking beyond the war years, this study illustrates significant continuity in the nature and timbre of hydro regulation across the first half of the twentieth century. Such regulation was certainly not "lax" or "weak" as hydro's original historians have suggested;[11] as I have shown elsewhere, it was actually quite well-intended and even rigorous, at least to the extent that small budgets, as well as larger political exigencies, allowed.[12] The original Hydraulic Service chief, Arthur Amos (who served between 1910 and 1937), was certainly a pro-development Liberal in the mould of the Gouin and Taschereau eras, but he was also fiercely protective of Quebec's hydraulic domain, and doggedly determined to exact financial remuneration for its exploitation by private interests. So was the young man he hired in the 1920s, Raymond Latreille. Born in Montreal in 1898, Latreille attended that city's distinguished and classical Collège Mont-Saint-Louis and then Canada's pre-eminent French-language engineering school (as did Amos), l'École Polytechnique de Montréal. He graduated with a degree in civil engineering in 1922. Three years later he joined the Hydraulic Service, and in October 1940 he assumed the position of chief.[13] Latreille was named by Premier Godbout as one of the four original commissioners of Hydro-Québec who took possession of the offices of Montreal Light, Heat and Power in April 1944. He would also oversee the completion of nationalization in the early 1960s (the only original member of the Commission to do so) before retiring in 1966. Some of the historical roots of Hydro-Québec are found in the efforts of civil servants, such as Amos and Latreille, to publicly regulate the private production and distribution of electricity. We might note, however, that the memoranda and decisions of interim Hydraulic Service chief Normandin lack something of the clarity and metal of the other two. That Duplessis sacked Amos in 1937 but kept Normandin on, and then re-engaged the latter as hydro adviser after re-taking power in 1944 perhaps reflects the fact that Normandin lacked the backbone or nationalist grit of an Amos or a Latreille. As shown elsewhere, hydro's civil service remained beholden and vulnerable to elected political authority in this early twentieth-century period of waterpower regulation.

The political concession process with regard to the Saguenay during the entire 1900 to 1940s period helps us to understand which constituents participated in decisions about watershed use. Throughout their careers, Hydraulic Service chiefs Amos and Latreille negotiated in earnest with private companies and investors in the hydro, forestry, and mining sectors. Their intent, always, was to ensure that Quebec's waterpowers were developed in the best economic interests of the province. They also looked out for the interests of less influential constituents: French-Canadian citizens, whether farmers or factory workers, whose livelihoods might benefit (or suffer) from industrial growth. Note the state's relentless emphasis through the Peribonka negotiations on gathering provincial revenues and promoting industrial development; in other words, doing what it could, short of government ownership, to win province-wide compensation for the concession of the Crown's precious hydraulic resources. As Godbout himself understood, to place a dam in a river was "mettre en valeur": to garner use and economic value from what otherwise ran wastefully to the sea.[14]

What we today would call 'environmental' concerns did not play a role in hydro decision-making: fishing denoted tourism, but fish themselves were all but irrelevant to the political process. Moreover, Natives, who served as guides to Alcan's survey crews and who trapped the lands where subarctic reservoirs were built, certainly did not yet hold a stake in the resource politics of Quebec. Aboriginals and their interests are entirely absent from the historical record in the matter of lakes Manouan or Passe Dangereuse, or the rivers of the North Shore; they were neither consulted nor compensated. Phrased another way, Native Peoples and their interests are everywhere *present* in the historical record by their omission: the government of Quebec could permit Alcan to expand high in the boreal uplands of the Peribonka basin precisely because it was 'merely Indians,' a relatively powerless and disenfranchised minority,[15] that stood in the way. Furthermore, Amos and Latreille would likely have pointed out: Natives held no treaty or title to the land in question. Nor were Natives seeking to exploit the land in a manner to benefit a broader citizenry at large, as the state was doing; they lacked both the means and the inclination to do so. In the estimates of Amos and Latreille, therefore, the interests of the many held clear precedence over the interests of the few. That the 'many' were French Canadians would have been highly significant to these nationalist civil servants who were so determined to advance the cause of their race, of "notre peuple," as Latreille put it.[16] How much more useful to place dams in Quebec's rivers, which yielded factories on their shores, which in turn paid wages to French-Canadian workers. How much more valuable to garner revenues for the provincial coffers while simultaneously elevating living standards for a *Canadien* majority long relegated to second-class economic status in its own homeland.

As participants in the fur trade, the Innu of this region had retained a degree of independence and autonomy, as has been documented in these pages. Through the 1930s, they exercised the authority to decide where and how to practise the rigorous art of trapping and hunting; they sustained themselves largely from the animals, plants, and trees of the northern forest; and by taking advantage of the opportunity to trade

their pelts to traders other than those of the HBC, they held a measure
of leverage over price and credit in the trade itself. Yet such microeco-
nomic influence on the mechanics of the fur trade and the subsistence
of the hunt in no way guaranteed the Montagnais a seat at a much
more important bargaining table: in Quebec City, where civil servants,
politicians, and industrialists, responding to wartime's surging economic
demand, redrew the map of the subarctic. In conceding to Alcan the
rights to fashion broad reservoirs at the headwaters of the Peribonka,
the Quebec state replaced the fur trade with the machine-driven extraction of wood, ore, and energy, rejecting an old staple commodity in
favour of more profitable new ones. Even the venerable Hudson's Bay
Company, let alone the Innu trappers with whom it did business, was
dismissed from this negotiating process. In this wartime-era contest of
'tribe' versus 'tribe' (with still decades to go before the brouhaha over
the development of James Bay), there was really no contest. The Quebec
state would see to it that the interests of French-speaking Quebecers
were honoured above that of others. The presence of a 'Native voice'
in Quebec hydropolitics would await the 1970s and the era of Native
Claims and North American environmentalism.

The effect of the new storage reservoirs on the Montagnais's hunting
society was immediate and direct, due to both the destruction of animal
habitats and the degradation of major waterways as natural highways of
transportation.[17] The disturbance of natural fluctuation in water levels
in the reservoirs and along the major rivers controlled by the dams
rendered these environments inhospitable to some animal and fish
populations. Beaver, in particular, cannot survive in reservoirs that are
drastically drawn down in winter; it leaves the animals vulnerable to
wolves. Lake trout, too, diminished in the reservoirs, and ouananiche
disappeared altogether in the rivers below as their spawning grounds
were disturbed by irregular water flows and eventual (future) timber drives. Muskrat were also much affected; for example, the Passe
Dangerouse reservoir flooded an expansive and watery grassland known
for being rich in muskrat. High river flow in the normally low-water
months of winter rendered the Manouan and Peribonka rivers dangerous for travel by sled or snowshoe; similary, large and wind-blown
reservoirs bearing floating dead wood hindered safe canoe travel in
the spring, summer, and fall. As meat and fish disappeared from the
hunter's path, the annual Innu journeys upstream and downstream
became increasingly impractical. Of course, these 'push' factors on
a semi-nomadic people to vacate their forest homes and settle down

on the reserve were accompanied by the 'pull' of Federal government
social welfare policies in the years immediately after the Second World
War. Family allowances contingent upon mandatory school attendance
increasingly drew mothers and their young to the village and removed
children from the necessary apprenticeship of bush life; Ottawa built
permanent homes for these new inhabitants in Pointe Bleue and
Bersimis. Government intervention was also evident in a so-called beaver preserve, established by the Department of Indian Affairs in the
Peribonka basin in 1941, which placed restrictions on Natives' trapping
and produced considerable resentment and lack of co-operation among
the trappers themselves. This was a markedly different response from
that of the Cree of the James Bay region who worked collaboratively
with provincial and federal officials to implement a similar conservation
program.[18]

How dam construction wrought social change in the long term is
equally noteworthy. Passe Dangereuse's construction necessitated a
lengthy access road, as we have seen. This road and its many offshoots
offered whites a point of entry to an expansive territory that had long
been barred to them by the impassable rapids of the Peribonka River.
"Before there was a dam there, they couldn't go up 'the long rapids'
[Passe Dangereuse]," Innu Gerard Siméon explained. "It was after the
flooding that the Whites began to go up."[19] In the wake of the Second
World War, forestry operations rolled northward into this new territory
as the provincial government issued new timber limits. Hunting and
fishing clubs as well as hunters' *chalets* proliferated in this same period
with the sanction of provincial government leases; several dozen clubs
were established throughout the hunting territories of the Pointe-Bleue
and Bersimis bands by the 1970s.[20] Some of the fish and game clubs
employed Native guides, but most also barred Native hunting and trapping. The white sportsmen that such clubs attracted killed for recreation
rather than subsistence. Many whites hunted without guide or licence;
in other words, they poached. Aluminum Company personnel witnessed
this rapid incursion into the "vast, virgin territories" of the Saguenay's
uplands. "The construction of the road and [then] forestry operations
opened the territory to poaching and extensive hunting," Arvida's Omer
Bernier recounted to McNeely Dubose in 1952, a decade after Passe
Dangereuse's construction. "During the hunting season last fall, I went
up past the forks of the Manouan," described Arvida's Yvon Cousineau.
"There were a lot of people there – more than there are usually seen in
the streets of Arvida on busy days. Judging by the number of people …

and the poaching that is taking place, in two or three years there will be no moose left." Summarized Cousineau: "If a district is rendered more accessible, the number of animals there is bound to be diminished."[21]

In short, building the dam required a road, and the road flung open to industrial and recreational use a vast boreal upland hitherto exclusively used by Native hunter-trappers. The consequence was the large-scale disturbance of animal habitat and the eviction of indigenous populations from ancestral territories. Natives who insisted on pursuing their trapping in the region often found their caches vandalized or traps stolen. "The wolverine breaks everything: caches, tents. And it pisses on the provisions," said Pessamit hunter Joseph Bellefleur. "Today, the White Man is like the wolverine."[22] Innu hunters, meanwhile, observed a precipitous decline in the population of moose, caribou, and various fish and furbearers through the industrial 'boom' of the 1950s and 1960s, to the point that hunting and trapping became impractical as full-time occupations. After all, even subtle changes to the boreal ecosystem rendered subsistence in this hard winter environment untenable. So in the post-war decades, beginning in the 1950s, it became increasingly difficult, and eventually impossible, to sustain oneself and one's family from the bounty of the land. "When the White Man touches something, he destroys it," Innu hunter Jean-Baptiste Dominique put it bitterly. "It's only the rocks that the White Man cannot destroy."[23] Of seventeen family hunting territories in the Peribonka basin, just two were actively being trapped and hunted by Montagnais by the middle of the 1970s. In any case, new employment opportunities as hunting or fishing guides, or as log drivers, tempted Innu hunter-trappers with alternative livelihoods.[24]

In ways both direct and indirect, large-scale hydroelectric projects of the war years (and beyond) helped precipitate a social revolution among the Innu/Montagnais of the Saguenay watershed and Quebec's North Shore generally. The Peribonka's lacs Manouan and Passe Dangereuse projects inaugurated a veritable frenzy of wartime and post-war storage and power dam construction on the rivers of the Saguenay-Côte Nord region, including on the Peribonka (1940, 1942, 1952, 1953, 1960), the Bersimis (1956, 1959, 1961), the aux Outardes (1969, 1969, 1969, 1978), and the Manicouagan (1958, 1960, 1965, 1967, 1967, 1972, 1972, 1976). The Second World War marked a historical watershed for the Innu of Quebec's interior and other Native groups of the Canadian Shield[25]: a chronological line of divide between their relative isolation and self-sufficiency and the full-scale invasion by Canadian civilization.

In the decades following the war, sedentary life on the reserve, a life largely dependent on government assistance, came to supplant the semi-nomadism of the previous millennia.[26]

Those trappers who experienced this dramatic transition, a number of whom had helped to ease it forward by working for Aluminum Company, harboured considerable regret and bitterness in later years. Jack Germain became Pointe-Bleue band chief in the 1930s and personally led Alcan's hydraulic engineers to what became the sites of both the Lac Manouan and Lac Peribonka (Passe Dangereuse) reservoirs. For this service, he remained on the company payroll for nine years; he also earned a footnote of acknowledgment in Paul Clark's company history, *Rivers of Aluminum,* a manuscript that was never published: "Tribute must be paid to [Fred Lawton and Charles Miller's] guide, Jack Germain, a full-blooded Montagnais with remarkable power of observation and retention and extraordinary physical strength," wrote Clark in 1964. "He had trapped the vast land for half a century to know more about thousands of square miles than the cityman can relate of his own street. Jack's ability to point to a map and state elevations, depths of lakes and rates of run-off saved time and when instruments checked his appraisals, they proved him astonishingly close to the mark."[27] Germain would come to regret the role he had played. Regarding the Passe Dangereuse reservoir, he told Gilbert Courtois in 1977: "Lac Peribonka hurt all the hunters. It did damage … There is nothing left … There were muskrat … but today not one. There are none, they have nothing to eat. It's all flooded there now. There's only rotted wood."[28]

We should explain that while Germain was the most important Native employee of Alcan, he was hardly the only one, as is revealed in Hudson's Bay Company post journals. The HBC post at Pointe-Bleue evidently acted as supplier to both the Saguenay Power Company's survey crews and the Native-hiring agent for Aluminum Company. As of 29 June 1940, Pointe-Bleue manager H.B. Frankland was filling a supply order for "Mr [Charles] Miller of Saguenay Power who intends damming the upper Peribonka river around Lake Manouan. Jack and Bartlemy Germain taken as guides and a fair order to be supplied next week." For 11 July we read: "Busy with Preliminary order of $100 for Saguenay Power, in connection with dam being constructed at Lake Manouan on upper reaches of Peribonka. Jack and Bartlemy Germain employed as guides to find winter road for tractors." For 25 July: "Busy with Saguenay Power business, order today amounts to $350, and 2 more guides given jobs with them." As Alcan extended its surveys to the Passe Dangereuse site that autumn, additional notations appear. On 22 October: "Aeroplane busy overhead taking supplies to a survey party on the Peribonca River on account of the Saguenay Power." 2 November: "Some goods sent into Saguenay Power and employees working on Lake Peribonca via Dominion skyways." 18 November: "Occupied with Mr Dechman of Saguenay Power organizing trip by dogs into Lake Peribonca. David Philippe and Willie Nepton given this job for the winter." And for 29 January 1941: "J.B. Germain took the plane from Roberval to join the [survey] gang working at Lake Peribonka." By 10 February, the post manager was "visiting around village [of Point Blue] and selected four men to report [to] work [at] Manouan dam on 4 month contract." Other Natives, presumably, were hired on directly in the bush. 13 March: "Manager busy in the office as Manouan paychecks arrived; amongst collections [is] current account of Willie Simpson and $20 on old account." The Bersimis post agents also noted the unusual activity of its Native customers when, in December 1940, relatively few Montagnais returned from the woods for the Christmas holiday season. "It would seem that *quite a few* [italics added] of the hunters from the Bersimis River are working at the construction of Lake Maniwan [*sic*], thus there is likely to be very little hunting done in that region this year."[29]

While the phrase "quite a few" hardly provides us with a specific number, it does suggest a significant level of Native participation in dam construction. Most would have likely been woodcutters or common labourers, rather than guides. And we should note that as their work was arranged with the knowledge and even co-operation of HBC, these HBC agents were, ironically, helping to sow the seeds of the fur trade's destruction in the Saguenay watershed and along the North Shore. So, of course, were the Aboriginals themselves. As Frank Tough has written about the result of Native participation in wage labour in Northern Manitoba: "their material life improved over that of the old fur trade regime, but their lack of control over land and their role as labourers meant that they were not in a position to secure long-term benefits." Thus, like any nascent working class in a period of industrialization, we must consider that "Natives actively participated in historical processes that contributed to the inequality of the present."[30]

Like Germain, hunter Gérard Siméon worked for the company; and like Germain, Siméon later expressed regret for what had resulted. "On the river itself, you can't live there. There are too many problems. There's the water that rises and falls, you know, it changes all the time. You can't tend traps there. It's ruined." Siméon did manual labour at the Passe Dangereuse construction site with his father and brother. But he could not have imagined the consequences such dam building would hold for the future. "Yeah, I worked on the construction of that dam there at the Passes ... But back then, you know, we had no idea, you know, that if they built a dam like that ... If we had known it could do harm, damage things, that we were destroying what belonged to us ... [but] no one spoke up, no one said a word."[31] Siméon mused more softly in his old age on the demise of his own hunting society by the inexorable encroachment of industrialization: "We were a kind of wild flower. When our sun and water and nourishment were withdrawn, we could no longer survive."[32]

How and whether to assign blame for this quiet human tragedy remains an open question. What the Montagnais experienced was typical of those that dwell in resource regions whose development is commanded from afar. Anthropologist Paul Charest has shown how the exploitation of natural resources of Quebec's North Shore (from fish and fur to lumber, paper, and iron ore, across five centuries) was directed by outsiders, at the expense and to the detriment of local people.[33] What was true of the North Shore has long been true of Canada's Shield region generally. Writing perceptively of the early twentieth century, historian José Igartua has said: "The resource regions of Canada were simple pawns on the continental, even global chessboard of American corporate capitalism."[34] This story of the Peribonka watershed's industrial transformation conforms in a measure to this pattern. It illustrates the manner by which local, indigenous people were both excluded from crucial decision-making and dispossessed of the land as a result and the manner by which the United States, including a Canadian subsidiary of an American corporation, and the bureaucracy of the national government in Washington, DC, provided a principal motive for the Peribonka's exploitation. Yet tracing the derivation of lakes Manouan and Peribonka proves more complicated still. It was a European nation, Great Britain, besieged by the Nazi attack in 1940, whose dire needs for aluminum spurred the creation of Lac Manouan; and without Alcan's contribution of metal to the UK aircraft program, said E.C. Plowden, director general of materials production at the British Ministry of Aircraft Production, "We would have lost the war."[35] Both lakes Manouan and Peribonka were also formed under the influence of a provincial government determined to develop its resources for the larger benefit of French-Canadian society, thus, by the force of ethnic nationalism. So, too, did the government of Canada play a role, by prodding hydroelectric development to serve the larger aims of Canadian industrial growth and the vigorous prosecution of global war.

The Second World War itself, an outgrowth of factors far too deep and complicated to probe here, was perhaps the ultimate agent in accelerating industrial production on the North American resource frontier – and, in this way, stimulating encroachment on Aboriginal land. In his highly original study of the Canadian military and Aboriginal land, P. Whitney Lackenbauer has explored the use and appropriation of Indian reserve lands during the twentieth century for military training and/or defence installations, as well as the resulting impact on Native communities.[36] Our own study of the Peribonka valley reminds us that the impact of war on Native homelands extended well beyond strictly military use by the government of Canada, to the widespread mobilization of strategic resources by both private and public interests, which was carried out in response to continental and global demand. As Morris Zaslow generalized for the Canadian North as a whole, "the war left a much poorer environment for wildlife in large sections of the North, greatly reducing the prospects for a healthy post-war [hunting and trapping] industry."[37] As we probe further, the causes of large-scale resource exploitation – which, in turn, catalyzed the demise of a hunting culture – become increasingly multiple and collaborative and vague. The "death" of the Innu as a nomadic society, as Gerard Siméon understood this process late in his life, resulted from "many little fires" rather than one single factor or event.[38] Aboriginals were most certainly victimized, and their lands vandalized, in the upper Saguenay watershed; nevertheless, it becomes terribly complicated to fix blame on a particular perpetrator or phenomenon.

It is also instructive to compare the Innu's exclusion from decisions around hydroelectric development to what occurred with the Natives of western Canada in a somewhat later period. Anthropologist James Waldram, in his study of the hydroelectric development of the Churchill, Nelson, and Saskatchewan rivers from the late 1950s as affecting Native communities in Manitoba and Saskatchewan, found that the manipulative process of treaty- or scrip-making of the nineteenth century was reduxed by western provincial governments in the mid-twentieth century in order to remove the Natives' legal rights from the waterways in question. In turn, removing Aboriginals physically and bureaucratically from their water resources, so crucial for fishing and transportation, deepened their condition of poverty and dependence.[39] The Montagnais-Innu's experience with big hydro was, at least initially, even worse. They enjoyed no legal claims to the lands beyond the small and restrictive borders of the Pointe-Bleue or Bersimis reserves, since no treaties had yet been negotiated between the Province of Quebec and the Native peoples of the Middle or Far North.[40] Representatives of the province, such as Latreille, could feel no obligation to engage Natives in decisions around resource development and could safely disregard Aboriginal interests. Latreille understood that building reservoirs in Indian country, well removed from the interests of French-Canadian farmers at Lac St-Jean, "far from the centres of colonization," made good political calculus.

In the long term, ironically, the exclusion of the Innu from such decisions in the 1940s may yet serve their interests. Since the 1970s, some means of compensation for territorial dispossession, in the form of renewed jurisdiction over ancestral lands, has been under negotiation between the Innu-Montagnais and the governments of Quebec and of Canada.[41] Initial results, in the 2000 document called "Joint Approach"/"Approche Commune" as well as 2004's "Agreement-in-Principle"/"Entente de principe," suggest that the Innu will benefit from terms regarding land claims and self-government that are far more economically lucrative and politically favourable than they could possibly have obtained in earlier decades. Delaying political negotiations until the late twentieth century, by which time Canadians saw themselves as a multicultural and tolerant society, was certainly not a conscious strategy on the part of Quebec's Montagnais-Innu. But it may well prove to be an effective one.

Drawing conclusions about Aluminum Company of Canada's corporate strategy and/or geographic-economic decision-making is constrained by the sources we have been able to access. From the various government documents that form the centrepiece of this paper, Alcan emerges as a smart, aggressive, decisive, and, at times, wily negotiator, self-interested in enlarging its resource base in Quebec for the purpose of growing shareholders' profits. Corporate growth was clearly a geographic, in addition to financial, phenomenon[42]: Aluminum's comparative advantage in Quebec, from the 1920s forward, was cheap abundant power on or near the Saguenay's tidewater fjord. Wartime permitted the firm to extend its control to the water storage in the Peribonka and Manouan basins, although these efforts were parried by civil servants who insisted on leases, not sales, and were careful not to give away more than was necessary per project. That Aluminum's efforts to expand responded to larger continental and global phenomena – whether the economic cycles of the 1930s or the burgeoning demand of the Second World War for ingot – is a clear pattern through this narrative. That Aluminum enjoyed privileged political access should also be evident from this study: the historical record suggests at least five special meetings of the Quebec Cabinet (excluding the passage of Orders-in-Council) devoted exclusively to the demands of Alcan's executives, as well as several additional one-on-one meetings between Aluminum representatives and the premier.

Aluminum also closely coordinated its activities with Ottawa. In fact, Alcan's relationship to Canadian officials remained respectful, even cordial, despite Herbert Symington's pressure on the company to get on with the Shipshaw dam. It is not clear when Ray Powell and C.D. Howe (who shared an interest in fishing and golf, along with their American roots[43]) formed their friendship inside the small circle of Canada's business-political elite. By the summer of 1944, certainly, "Rip" and "Clarence" were discussing salmon trips on the Ste-Marguerite River, games at the Royal Ottawa, and weekends at Powell's "Bungalow."[44]

But providing nuance to our portrait of a powerful and well-connected multinational corporation is difficult without additional documentation. We have pored through official histories and company memoirs, and used the unscripted data, commentary, or testimony that is preserved in government archives. Missing here are the internal corporate memoranda and correspondence and Minutes to which Alcan has repeatedly denied this author (and others) access. Such material, including the 'presidential correspondence,' bound in blue volumes and held tightly by Alcan's Business Information Centre, might well speak to a whole batch of unanswered questions that are raised in this study. How did Quebec's changing political climate, specifically the rising winds of hydro-nationalism, shape or check the company's desires to expand its access to the province's hydro-rich watersheds? How did Ray Powell and McNeely Dubose view the respective administrations with whom they bargained? or the state's engineer-negotiators Normandin and Latreille? or competing constituencies such as Big Timber? In the eyes of corporate executives, was the government merely a bureaucratic obstacle to be cleared, or did company executives also have some sense of patriotic responsibility toward the war effort (as the official history claims and others have since echoed[45]) and/or the provincial and Canadian economies? Was Aluminum truly as reluctant to build Shipshaw as Herbert Symington reported, or might Alcan's executives have tapped Ottawa's strong motivation to grow Canada's energy arsenal in order to win better federal tax breaks? Did the company in fact seek to exploit the crisis of wartime to enlarge its resource base in Quebec, as Latreille and others suspected? The writing of business history is so often limited by the lack of an insider's knowledge or documentation. Unfortunately, this is the case here.

What is clear is that lakes Manouan and Passe Dangereuse/Peribonka served the company's interests in the long term, as new storage capacity rendered the construction of additional hydroelectric dams downstream practical and profitable. The Second World War ushered in the 'aluminum age.' The war produced a five-fold increase in North American aluminum smelter capacity, which made the light and flexible metal readily available for public transportation, housing, automobiles, and a host of new consumer goods. By the end of 1946, Alcan began to benefit from such new market niches. Surging post-war demand for power for paper as well as ingot manufacture moved Aluminum Company to once again seek permission from the Quebec government to divert the waters of the Bersimis River. Refused a second time, the company instead sought and won permission from the Duplessis government in 1950 to construct two additional power dams on the lower Peribonka at sites long identified by the Quebec Streams Commission: Chute-du-Diable and Chute-à-la-Savane. Completed by 1953, with each dam amassing 110 feet of the river's fall, the power plants added some 410,000 installed horsepower to the Saguenay system, bringing the company's total installed capacity in the region to 2.5 million horsepower. Further consumer and industrial growth through the 1950s spurred yet another major hydroelectric project in the gorge below Passe Dangereuse. Dubbed 'Chute-des-Passes' and constructed between 1956 and 1960, the power plant (an ambitious structure built deep underground at the terminus of six-mile-long intake tunnels dug parallel to the gorge) added another million horsepower to the Saguenay power system. To better serve this plant with water, Alcan shut down the natural flow of the upper Manouan River to canalize Lac Manouan's waters westward to the Peribonka and thus through the turbines of Chute-des-Passes.[46]

Most recently, Hydro-Québec set out to use the storage potential of lakes Manouan and Passe Dangereuse by constructing a hydroelectric power plant at the Manouan-Peribonka forks, where Innu hunter-trappers once assembled on their annual migrations to and from Lac St-Jean. Completed in 2008 at a cost of $1.2 billion, the eighty-metre-tall dam plus Péribonka IV powerhouse adds 385 megawatts (some 516,000 horsepower) of generating capacity to the watershed. As befitting current political trends and sensibilities, Hydro-Québec's work at the Peribonka was accompanied by human and environmental impact studies, archeological digs (in association with Université du Québec à Chicoutimi and the Innu of Mashteuiatsh), a compensatory agreement with the Innu, and a website that followed construction progress via animated diagrams, colour photographs, and live webcam.[47] How the politics of development have changed in the course of a half-century![48]

Wartime laid the hydraulic foundations for the Peribonka's future exploitation. Strategic and consumer demand of the Cold War and post–Cold War periods continue to reshape the river for hydroelectric production. The Peribonka, of course, is but one of dozens of Canadian Shield rivers to have been tapped in the post-war era. In the largest sense, then, the Peribonka concessions exemplify the history of an environment, as well as the environmental history of war. Our unit of analysis is the watershed, and our major concern is the decisions that reshaped this landscape for industrial use. Clearly, this is an environment that has been shaped from afar: Canada's subarctic or 'Middle North' has long dwelt in the economic orbit of the industrial heartland to the south. Across the twentieth century, the burgeoning consumption of pulpwood and minerals and energy by North American urban populations motivated governments and businesses alike to encroach ever northward into the Canadian Shield. Such is the historical characterization of one of the pioneers of Canadian Northern history, Morris Zaslow. Our study of the Peribonka concessions certainly echoes and pointedly illustrates Zaslow's claims.[49]

Twentieth-century warfare certainly accelerated this process of environmental change.[50] During the First World War, American financier James Duke, in collaboration with the E.I. du Pont de Nemours Powder Company (Du Pont) of Wilmington, Delaware, had sought to convert the Saguenay's tremendous energy potential into factories producing nitric acid, the key ingredient of high explosives. In that earlier era, it was government that hampered the grand industrial plan by throwing up regulatory obstacles, before the war itself made the market for both dollars and labour too tight to proceed.[51] Actual development of the Saguenay awaited the interest of Alcoa in the 1920s. In the 1940s, our evidence suggests that it was government, rather than capital, that spearheaded large-scale hydroelectric development, while a reluctant private corporation, Alcan, eventually bowed to Ottawa's bidding. And it was nation-states, in multi-state alliances, which mobilized large quantities of hinterland resources to fashion the tools and weaponry of global conflict. Aluminum Company of Canada, called on to provide as much as one-third of the Allies' ingot requirements during the war, was a key contributor to this effort. Canada's provincial and national governments eagerly facilitated this arrangement, even while attempting to use the war's crisis to secure future industrial development and a measure of financial remuneration for their constituents. The result was the concession of two broad reservoirs in the Peribonka valley. In this scenario, private companies and governments, alike, played but supporting roles in the larger global drama of war. Ultimately, it was war that remade waterpower at the Peribonka River. In the global conflagration of the 1940s, Canada's northland served as but one of the war's subarctic fronts.

Still, the war didn't simply accelerate resource development in this case; it also checked its pace and changed its nature. Wartime's frenzied tempo of economic mobilization did hurry, and weaken, decisions to permit the creation of lakes Manouan and Passe Dangereuse. But so, too, did rapid wartime expansion by Alcan encourage Quebec's functionaries to place limits on the firm's territorial growth. In refusing the company concessions beyond the Saguenay watershed, the government preserved the valleys of the Bersimis, Manicouagan, and aux Outardes for later exploitation by the state itself. Similarly, the political backlash to rapid mobilization by Aluminum Company in the province seemed to hasten the decision by politicians to assert further government control over hydraulic resources by nationalizing electricity.[52] In this sense, the vast reservoirs of the upper Peribonka River that are the object of this study form only part of the enduring environmental legacy of the Second World War. Far more significant in the long term was the founding of Hydro-Québec. Its energetic, post-war, dam-building efforts in the service of French-Canadian economic advancement – on the North Shore, at James Bay and Hudson's Bay, and elsewhere – has created the world's largest producer of hydroelectric power, has instilled in many Quebecers a sense of pride and accomplishment, and has laid an economic foundation for any future movement toward political sovereignty. Such were the unintended human-environmental outcomes that flowed from the Saguenay basin's wartime hydraulic history.

Notes

ABBREVIATIONS AND SHORTENED REFERENCES

Alcan: Alcan Aluminium Limited, Montreal, renamed Rio Tinto Alcan in 2007.

BANQQ: Bibliothèque et Archives nationales du Québec, Quebec City branch.

MAM: Musée amérindien de Mashteuiatsh, Mashteuiatsh/Pointe-Bleue, Quebec.

CIP: Conseil des Innus de Pessamit, Pessamit/Betsiamites, Quebec.

LAC: Library and Archives Canada, Ottawa.

RN: Ministère des Ressources naturelles et de la Faune, Charlesbourg, Quebec.

QSP: Quebec Sessional Papers.

WPB: War Production Board, National Archives and Records Administration of the United States, College Park, Maryland.

INTRODUCTION

1 Any historian setting out to write about Canada's Native Peoples must deal with the politics of nomenclature. Herein I will use what seem to be the most descriptive and least controversial terms: 'Aboriginals' and 'Natives,' as well as the more specific 'Montagnais' or 'Innu' for the people of Quebec's eastern interior. I have chosen to avoid the potentially offensive term 'Indian,' despite its widespread use across Canada, including among Aboriginals themselves. (Exceptions are made when quoting historical documents or describing proper nouns such as Point Blue Indian Reserve or Department of Indian Affairs.) I have also eschewed the more politically correct 'First Nations' as hopelessly anachronistic, i.e., implying a degree of governmental and territorial

organization that did not exist among Native groups when Europeans
arrived in North America.

2 Informateur 73 (Louis-Georges Boivin), Fiches de description du site
de campement for the voyage of 1938, MAM. I have simplified the
chronology somewhat to condense it into three short paragraphs.

3 James Scott, *Seeing Like a State* (New Haven and London: Yale University
Press, 1998).

4 This is not a study in labour history, for several reasons. As a study of
landscape origins, this book stops where shovels strike dirt or axes fell
trees, so it does not encompass the dam-construction process itself.
Second, the workers, most of them French Canadians, who built the
reservoir's dams were not directly involved in the state's lease negotiations
with Alcan (though we might note that the government sought to
negotiate in the interests of all Quebecers). Finally, the workers' story is,
to some extent, already dealt with by Alcan's company literature. See, for
example, Campbell's *Global Mission* and Clark's "Rivers of Aluminum", as
well as F.L. Lawton's "The Manouan and Passe Dangereuse Water Storage
Developments," in *The Engineering Journal*, 27, 4 (April 1944).

5 The resulting controversy is dealt with in Victor Tremblay, *La Tragédie du
lac St-Jean* (Chicoutimi, QC: Editions Science Moderne, 1979).

6 'Peribonka' is the traditional English-language spelling and was used by
both industrialists and bureaucrats in the early twentieth century when
the events of this study take place. The French spellings, 'Péribonca'
or 'Péribonka,' are both commonly used in modern Quebec. All, of
course, are but transliterations of the original Algonquin term. Other
place names have been standardized in the text, including 'Manouan'
(occasionally spelled 'Manouane' in the sources) and 'Passe Dangereuse'
(sometimes written in the sources as Passes-Dangereuses).

7 See Paul Charest, "Les barrages hydro-électriques en territoire
Montagnais et leurs effets sur les communautés amérindiennes,"
Recherches amérindiennes au Québec IX, 4 (1980): 323–37, which utilized
interviews conducted by Gilbert Courtois in his *Étude sur l'impact des
barrages hydroélectrique* (unpublished, 1978), as well as Raphaël Picard's
*Impact socio-économique des barrages hydro-électrique des rivières Bersimis, aux
Outardes, Manicouagan sur les Montagnais de la bande de Bersimis* (Rapport
d'entrevues remis au Conseil Attikemek-Montagnais, unpublished, 1978).

8 B.J. McGuire and H.E. Freeman, "How the Saguenay River Serves
Canada," *Canadian Geographical Journal* 35, 5 (1947): 200–25. Useful with
regard to c.f.s. is Aluminum Power Co. to Public Service Board, 17 August
1942, Contract file 2814, Alcan, Montreal.

9   David Massell, "'As though there was no boundary': The Shipshaw Project
    and Continental Integration," *The American Review of Canadian Studies*
    (Summer 2004): 187–222.

10 Nappi, "Canada: An Expanding Industry," in *The World Aluminum Industry
    in a Changing Energy Era*, ed. Merton J. Peck (Washington: Resources for
    the Future, 1988), 175–221.

11 Riotintoalcan.com. Accessed June 2008.

12 Camil Girard and Normand Perron, *Histoire du Saguenay-Lac-St-Jean*
    (Québec: Institut québécois de recherche sur la culture, 1995 (1989)),
    469–80; and Gérard Bouchard, "Le Peuplement blanc," in C. Pouyez,
    Y. Lavoie, et al., *Les Saguenayens: Introduction à l'histoire des populations du
    Saguenay XVIe–XXE siècles* (Sillery: Presses de l'Université du Québec,
    1983), chap. 4. On Alcan, see also Duncan C. Campbell's official
    corporate history, *Global Mission: The Story of Alcan*, 3 volumes (Montreal:
    Alcan Aluminium Limited, 1985, 1990, 1990); brief coverage of wartime
    construction of the Peribonka reservoirs is found in vol. 1, chap. VII,
    "The Second World War," 264–69. Most recently, historian Jean Martin
    describes the Second World War's positive effect on Alcan's growth in
    "Le Triomphe Canadien de la Deuxième Guerre Mondiale: L'Émergence
    d'Alcan dans l'Industrie Nordaméricaine de l'Aluminium" (unpublished
    paper, summer 2004).

13 Claude Bellavance, *Shawinigan Water and Power 1898–1963: formation et
    déclin d'un groupe industriel au Québec* (Montréal: Les Éditions du Boréal,
    1994), chap. 1.

14 In 1962 Jean Lesage's Liberals called a snap election on the issue of
    full-scale nationalization of the province's remaining private electrical
    utilities, with the goal of becoming, as the Liberals evocatively put it,
    "maîtres chez nous." Victorious, the Liberal government authorized
    Hydro-Québec to expropriate eleven private power companies (all but
    Alcan) at a cost of $600 million, half of this amount to be raised on
    Wall Street. With expropriation carried out in 1963, a much-enlarged
    and highly influential Hydro-Québec was born: an instrument of job
    creation and state growth during the Quiet Revolution, and a potent
    symbol of rising Francophone influence. Not surprisingly, this prominent
    institution has drawn adulation and criticism, and certainly considerable
    analysis, ever since. Among other works, see Clarence Hogue, André
    Bolduc, and Daniel Larouche, *Québec: un siècle d'électricité* (Montreal:
    Libre expression, 1979, 1984, 1989); Alain Chanlat, André Bolduc, and
    Daniel Larouche, *Gestion et culture d'entreprise: le cheminement d'Hydro-Québec*
    (Montreal: Québec-Amérique, 1984); Philippe Faucher and Johanne

Bergeron, *Hydro-Québec, la société de l'heure de pointe* (Montreal: Les Presses de l'Université de Montréal, 1986); Marcel Couture, dir., *Hydro-Québec: des premiers défis à l'aube de l'an 2000* (Montreal: Libre Expression, 1984); Taïeb Hafsi and Christiane Demers, *Le changement radical dans les organisations complexes: le cas d'Hydro-Québec* (Montreal: Gaétan Morin, 1989); Yves Bélanger and Robert Comeau, dir., *Hydro-Québec: Autres temps, autres défis* (Québec: Presses de l'Université du Québec, 1995); and Dominique Perron, *Le nouveau roman de l'énergie nationale: analyse des discours promotionnels d'Hydro-Québec de 1964 à 1997* (Calgary: University of Calgary Press, 2006).

15 Launched in the early 1970s, and continuing through the present day, the 'epic' of James Bay has excited both lavish praise and stinging criticism among journalists, economists, anthropologists, historians, geographers, and others for its sizable economic impact as well as its transformative effect on Aboriginal land and culture. Among much else, note the work of journalists Boyce Richardson, *Strangers Devour the Land* (Toronto: MacMillan, 1975); Roger Lacasse, *Baie James, une épopée: L'extraordinaire aventure des derniers des pionniers* (Montreal: Libre Expression, 1983); and the more balanced contribution, Sean McCutcheon, *Electric Rivers: The Story of the James Bay Project* (Montreal: Black Rose Books, 1991). More scholarly is Harvey Feit, "The Future of Hunters in Nation-States: Anthropology and the James Bay Cree," in *Politics and History in Band Societies*, eds. Eleanor Leacock and Richard Lee (Cambridge: Cambridge University Press, 1982); Feit's "Hunting and the Quest for Power: The James Bay Cree and Whitemen in the Twentieth Century," in *Native Peoples: The Canadian Experience*, eds. R. Bruce Morrison and C. Roderick Wilson (Toronto: McClelland and Stewart, 1986); and Richard Salisbury, *A Homeland for the Cree: Regional Development in James Bay, 1971–1981* (Montreal & Kingston: McGill-Queen's University Press, 1986). Recent additions to this literature include work by Caroline Desbiens, for example, "Nation-to-Nation: defining new structures of development in Northern Québec," *Economic Geography* 80, 4 (2004): 351–66; Hans Carlson, *Home is the Hunter: The James Bay Cree and Their Land* (Vancouver: University of British Columbia Press, 2008); and Thibault Martin and Steven M. Hoffman, eds., *Power Struggles: Hydro Development and First Nations in Manitoba and Quebec* (Winnipeg: University of Manitoba Press, 2008).

16 Most important is the work of Toronto-born, Harvard-trained economic historian John Dales (1920–2007) who may properly be considered the father of hydroelectric scholarship in Quebec,

and whose doctoral dissertation/book *Hydroelectricity and Industrial Development: Quebec 1898–1940* made the first geographical, historical, and statistical survey of the hydro industry. In the Innisian tradition of staples theory, Dales sought to understand how an industrially poor or 'backward' Quebec, uniquely endowed with waterpower, found its path to rapid industrialization. Specifically, he set out "to illuminate the relationship between hydroelectric growth and the growth of secondary manufacturing industry in the province." (pp. 1–2). Claude Bellavance, more recently, has explored Quebec's early hydro history from the vantage of a particular corporation in his excellent monograph *Shawinigan Water and Power 1898–1963*; while my own work (*Amassing Power*, 2000) looks at the business and politics of hydroelectricity through the political concessions of a particular watershed. Other works that deal tangentially or in part with Quebec's hydro industry of this era are: Christopher Armstrong and H.V. Nelles, *Monopoly's Moment: The Organization and Regulation of Canadian Utilities, 1830–1930* (Philadelphia: Temple University Press, 1986); Armstrong and Nelles, "Contrasting Development of the Hydro-Electric Industry in the Montreal and Toronto Regions, 1900–1930," *Journal of Canadian Studies* 18, 1 (1983): 5–27; T.D. Regehr, *The Beauharnois Scandal: A Story of Canadian Entrepreneurship and Politics* (Toronto: University of Toronto Press, 1990); Yves Roby, *Les Québécois et les investissements américains (1918–1929)* (Québec: Les Presses de L'Université Laval, 1976); Bernard Vigod, *Quebec Before Duplessis: The Political Career of Louis-Alexandre Taschereau* (Montreal & Kingston: McGill-Queen's University Press, 1986); and Pierre Lanthier, "L'industrie électrique entre l'entreprise privée et le secteur public. Le cas de deux provinces canadiennes: 1890–1930," in *Un siècle d'électricité dans le monde*, ed. F. Cardot (Paris: Presses Universitaires de France, 1986).

17 For the outlines of hydro policy during this period, see Claude Bellavance, "L'état, la 'houille blanche' et le grand capital: L'aliénation des ressources hydrauliques du domaine public québécois au début du XXe siècle," *Revue d'histoire de l'Amérique française* 51, 4 (1998): 487–520. *Amassing Power* offers a detailed case study of the 1890s–1920s period.

18 See Matthew Evendon, "La mobilisation des rivières et du fleuve pendant la Seconde Guerre mondiale: Québec et l'hydroélectricité, 1939–1945," *Revue d'Histoire de l'Amérique Française*, 60, 1–2 (été–automne 2006): 125–62; and David Massell's "'As though there was no boundary': The Shipshaw Project and Continental Integration," *The American Review of Canadian Studies* (Summer 2004): 187–222. Although neither deals squarely with Quebec policy or politics, both contribute to our knowledge

of the Canadian and continental contexts for wartime hydroelectric development in the province.

19 A fuller explanation of methodology is provided in Denis Brassard, *Occupation et utilisation du territoire par les Montagnais de Pointe-Bleue* (Village-des-Hurons: Conseil Attikamek-Montagnais, 1983), Annex A, Présentation de la recherche. Regarding the contents of the interviews, the intention was as follows: "Pour une année complete d'activités, le chasseur devait indiquer sur des cartes à l'échelle 1:250,000 tous ses itinéraires parcourus, sites de campement, portages, caches et lieux de sepulture. En plus des commentaires que soulève son récit, il devait pour chaque site de campement préciser les toponymes, la durée de séjour, les noms et nombre des individus et des familles installés au meme endroit et, s'ils existent, leurs liens de parenté. L'informateur devait aussi mentionner les activités pratiquées en ces lieux (chasse, pêche, trappe, cueillette) et les espèces récoltées. Enfin, il décrivait les installations de l'endroit (type, nombre et utilité) et commentait la progression du voyage. Pour s'assurer de la justesse de leurs propos, certains informateurs étaient assistés d'une personne qui avait fait partie du même groupe de chasse de l'année dont ils témoignaient. Ils vérifiaient ainsi leurs souvenirs et livraient un récit plus précis. Les informateurs ont parcouru les lieux si souvent et en campagnie de tant de gens que seuls des faits marquants permettent d'ordonner leurs souvenirs et d'avancer plus sûrement dans leurs propos: la naissance d'un enfant, le décès d'un proche ou un événement sans précédent marquaient souvent le depart de l'entrevue."

20 Ironically, I obtained full access to Alcan's holdings for the first segment of this study that dealt with Lac St-Jean (*Amassing Power*, Montreal & Kingston: McGill-Queen's University Press, 2000) only to discover that the vast majority of the files had been destroyed for the years before 1928 (when Aluminium Limited was created by Alcoa). But having spent several weeks in the archives in the sub-basement of Place Ville-Marie on the earlier project, I am aware of the so-called 'presidential correspondence' books dating back at least through the 1930s.

21 André Chevalier, Manager, Corporate Affairs to D. Massell, 16 October 2001.

22 I am drawing on Campbell, *Global Mission*, vol. 1; Charles C. Carr, *Alcoa: An American Enterprise* (New York: Rinehart and Company, Inc., 1952); and Isaiah A. Litvak and Christopher J. Maule, *Alcan Aluminium Limited, A Case Study*, Royal Commission on Corporate Concentration Study No. 13 (Ottawa: Minister of Supply and Services Canada, 1977).

CHAPTER 1

1 Pouyez, *Les Saguenayens*, chap. 4; Victor Tremblay, "La rivière Péribonka:
   les premières pages de son histoire," *Saguenayensia* 1, 6 (nov–déc 1959):
   143–6; V. Tremblay, "La rivière Péribonka: la période des explorations,"
   Saguenayensia (jan–fév 1960): 17–24; V. Tremblay, "La rivière Péribonka:
   période des chantiers et de la colonization, *Saguenayensia* (sept–oct
   1973): 134–62; Jeannette Girard and Jean-François Moreau, "Histoire
   et préhistoire de la rivière Péribonca," *Saguenayensia* 29, 1 (1987): 6–12;
   Russel Bouchard, "De Saint-Amédée à Chute-des-Passes: la colonisation
   de la Péribonca," *Saguenayensia* 37, 3/4 (1995): 3–23; Dany Côté, *De la
   terre, du bois, de l'eau et des gens: de Honfleur à Sainte-Monique, 1898–1998*
   (Sainte-Monique: Municipalité de Sainte-Monique, 1997).
2 The previous paragraphs summarize events that are described in more
   detail in Massell, *Amassing Power.*
3 Testimony of W.S. Lee, *Scott v. Quebec Development Company,* 55–6, Cour
   Supérieure, Québec, Archives nationales du Québec, Québec.
4 Léopold Naud Private Collection, Daily Journal 1914; and Naud
   Collection misc. documents dated 28 September 1915 and entitled
   "Rough Estimate of Cost of Making Borings at Proposed Dam Site on
   the Peribonka River, and Rough Estimate of Cost of Making Survey and
   Contour of Proposed Reservoir on the Peribonka River."
5 B.A. Scott to minister of lands and forests Jules Allard, 7 April 1914,
   RN, office of développement électrique, file Rivière Saguenay (Grande
   Décharge), Correspondence 1. Additional context for these negotiations
   is provided in Massell, *Amassing Power*, chap. 4.
6 Massell, *Amassing Power*, chap. 4.
7 Arthur Amos, Mémoire Re: Lac St-Jean et ses tributaries, 14 avril 1915,
   RN/Rivière Saguenay (Grande Décharge).
8 Gauvin's reports of 26 September and 10 October 1899 are in *QSP* 47,
   5 (1913); Langelier's "The Lake St. John Region with reference to the
   Pulp and Paper Industry," 22 February 1898, is in the Report of the
   Commissioner of Lands, Forests and Fisheries, *QSP* 32, 1 (1898–99).
9 Christian Pouyez, Yolande Lavoie, et al., *Les Saguenayens*, chaps. 2–3:
   "Les Amérindiens du Saguenay avant la colonization blanche" and "Les
   Amérindiens du Saguenay à l'époque contemporaine."
10 On Bersimis, a concise history is found in Pierre Frenette and Dorothée
   Picard, *Histoire et culture innues de Betsiamites* (Betsiamites: École
   Uashkaikan/Les Presses du Nord, 2002).

11 Fiche d'entrevue de l'informateur no. 16 (Jack Germain) and Fiches de description du site de campementfor the voyage of 1907, MAM.

12 The provision amounts are noted by anthropologist Frank Speck for the hunter Simon Raphael in/around 1915, but would roughly apply to the Germain voyage. Speck's work is found in "Family Hunting Territories of the Lake St. John Montagnais and Neighboring Bands," *Anthropos* 22 (1927), 387–403.

13 Fiche d'entrevue de l'informateur no. 39 (Patrick Étienne) and Fiches de description du site de campement for the voyage of 1914, MAM.

14 Speck, "Family Hunting Territories," 392.

15 Anne-Marie Siméon and Camil Girard, *Un monde autour de moi: Témoignage d'une Montagnaise* (Chicoutimi: Les éditions JCL, 1997).

16 J.W. Anderson, *Fur Trader's Story* (Toronto: Ryerson Press, 1961), 102–3.

17 Of the general histories or reports on Innu culture and life, as well as Innu-white relations, most useful to me have been the unpublished/internal studies of historical land use which were carefully prepared for the purpose of land claims negotiations: Denis Brassard, *Occupation et utilisation du territoire par les Montagnais de Pointe-Bleue* (Village-des-Hurons: Conseil Attikamek-Montagnais, 1983); Jacques Frenette, *Occupation et utilisation du territoire par les Montagnais de Betsiamites, 1920–1982* (Village-des-Hurons, Conseil Attikamek-Montagnais, 1983); and Jean-Guy Deschênes and Richard Dominique, *Nitasinan* (Village-des-Hurons: Conseil Attikamek-Montagnais, 1983). Also useful were Alain Beaulieu, "Du nomadisme aux reserves; histoire et culture des Montagnais du Québec," in *Les Indiens Montagnais du Québec: Entre deux mondes*, ed. Anne Vitart (Paris et Québec: Éditions Sépia, 1995), 11–33; Jean-Paul Lacasse, *Les Innus et le territoire* (Sillery: Éditions du Septentrion, 2004); Pierre Gill, *Les Montagnais, premiers habitants du Saguenay-Lac-St-Jean* (Pointe-Bleue: Mishinikan, 1987); René Boudreault, *Mashteuiatsh* (Wendake: Institut culturel et éducatif Montagnais, 1994); and Serge Jauvin, and Daniel Clément, ed. *Aitnanu: The Lives of Hélène and William-Mathieu Mark* (Montreal: Libre Expression, 1993). Montagnais history is set in the context of Native history in the province of Quebec in Daniel Francis, *A History of the Native Peoples of Quebec, 1760–1867* (Ottawa: Minister of Indian Affairs and Northern Development, 1983).

18 A useful introduction to the anthropological debate over the influence of the fur trade on Aboriginal land use is found in Eleanor Leacock, "The Montagnais 'Hunting Territory' and the Fur Trade," Memoir No. 78, *Memoirs of the American Anthropological Association*, 1954.

19 The Annual Reports of the Department of Indian Affairs contain, among other things, the annual reports of Indian agents for the respective reserves, including Point Blue and Bersimis, often containing information regarding population numbers and makeup, land under cultivation, and the like.

20 Reports on the "Montagnais of Lake St. John" in the Annual Report(s) of the Department of Indian Affairs during this period.

21 André Veilleux, "Pointe-Bleue: Histoire d'une reduction" (master's thesis, sociology, Université Laval, 1982), 87–8.

22 Reports on the "Bersimis Band" in the Annual Report(s) of the Department of Indian Affairs, 1906 and 1901. Similar comments are found throughout the reports of the pre-First World War period.

23 See Hélène Bédard, *Les Montagnais et la Réserve de Betsiamites, 1850–1900* (Québec: Institut québécois de recherche sur la culture, 1988), chaps. 3–4, especially pp. 78, 80. The fisheries officer and trapper Napoleon A. Comeau provides additional, if anecdotal, evidence of a turn-of-century Native-utilized interior in *Life and Sport on the North Shore of the Lower St. Lawrence and Gulf* (Quebec: Daily Telegraph Printing House, 1909).

24 This fact regarding the trade relationship is gleaned from a reading of the Hudson's Bay Company post records for both Point Blue and Bersimis, which constitute rich storehouses of historical information concerning the evolving nature of the fur trade in these locales.

25 Speck, "Family Hunting Territories," 389. Although Speck published his article in 1927, his information (as the author states) "pertains to conditions about 1912."

26 Anderson, *Fur Trader's Story*, 56, 58.

27 Siméon, *Un monde autour de moi*, 53.

28 Comeau, *Life and Sport*.

29 Conversation with Alain Nepton (Pointe-Bleue resident, hunter, former band council member, and consultant), 29 June 2006. In the post-war period, Nepton points out, the sense of place would gradually be inverted: the reserve would become 'home,' especially for women and children, and the bush a place for hunters' ever-briefer sojourns.

30 On timber harvesting in the Peribonka valley, see Girard and Perron, *Histoire du Saguenay*, 299ff.; and Dany Côté, *Riverbend: Splendeur et déclin d'une ville de compagnie* (Publication no. 8, Société d'histoire du Lac-Saint-Jean, 1994); D. Côté, *De la terre, du bois, de l'eau et des gens: De Honfleur à Sainte-Monique 1898–1998* (Municipalité de Sainte-Monique, 1997); and D. Côté, *Histoire de l'industrie forestière du Saguenay-Lac-Saint-Jean* (Publication no. 17, Société d'histoire du Lac-Saint-Jean, 1999).

Victor Tremblay describes the earliest-known foray for timber resources, conducted by surveyor Pierre-Alexis Tremblay on behalf of William Price, Sr beginning in 1854, in "La rivière Péribonka: La période des explorations." Massell notes Scott's activities in *Amassing Power*, 27–8, as does Carl Beaulieu in *B.A. Scott: Père de l'industrialisation* (Chicoutimi, QC: Les Éditions Entreprises, 1999), chap. 2.

31 Tremblay and Leclerc, Appendix 39, 91–4 in the Report of the Minister of Lands and Forests of the Province of Quebec, *QSP* 46, 2 (1912); Joncas, Appendix 41, 96–8, ibid; Jean Maltais, Appendix 29–30, 83–92, Report of the Minister of Lands and Forests of the Province of Quebec, *QSP* 47, 3 (1913). For a somewhat earlier account of the Peribonka region, see J.C. Langelier, "The Lake St. John Region with reference to the Pulp and Paper Industry," Report of the Commissioner of Lands, Forests and Fisheries, 1898, *QSP* 32, 1 (1898–1899), 80–95.

32 Jules Allard to Benjamin Scott, 16 April 1914, RN, Rivière Saguenay (Grande Décharge).

33 The third of Quebec's institutions pertaining to waterpower and electricity was the Public Utilities Commission, fashioned in 1909 to supervise electrical distribution and rates. While both the Hydraulic Service and Streams Commission seem to have done real work and carried real influence in the management of the hydro industry, this third organization has been criticized as toothless and ineffectual. See Christopher Armstrong and H.V. Nelles, "Contrasting Development of the Hydro-Electric Industry in the Montreal and Toronto Regions, 1900–1930," *Journal of Canadian Studies* 18, 1 (spring 1983): 5–27.

34 Massell, *Amassing Power*, chap. 4; Bellavance elaborates on the Commission's organization and function in "L'État, la 'houille blanche' et le grand capital", 501–6.

35 *Quatrième Rapport de la Commission des eaux courantes de Québec*, November 1915 (Québec: E.-E. Cinq-Mars, 1915), 70–84. Alcan's McNeely Dubose summarizes the Commission's work in "The Engineering History of Shipshaw," *The Engineering Journal* 27, 4 (April 1944): 194–9; 195.

36 By late 1916, the Streams Commission was slated to carry out a study of the Peribonka, as noted by Arthur Amos, *Les Forces Hydrauliques de la Province de Québec* (Québec: Département des Terres et Forêts, 1917), 45.

37 *Sixième Rapport de la Commission des Eaux Courantes de Québec*, 1917, 117; *Neuvième Rapport*, 1920, 10, 74–9; *Dixième Rapport*, 1921, 12, 15–19, 94.

38 Massell, *Amassing Power*, chaps. 6–7.

39 The annual Report(s) of the Minister of Lands and Forests for the period in question are found in *QSP*, then after 1935 as the semi-annual

Report of the Minister of Lands and Forests (Quebec: Redempti Paradis). Of particular use here have been the annual reports covering the operations of the surveys branch as well as the published extracts of the surveyors' reports themselves. See, for example, "Annual Report of the Director of Surveys," Appendix 19 of the *Report of the Minister of Lands and Forests of the Province of Quebec* (Quebec: Redempti Paradis, 1940), 147–54.

40 Report(s) of the Minister of Lands and Forests of the Province of Quebec, found in the QSP, vol. 61, pt. 1, 1928, 76; vol. 63, pt. 1, 1920, 89–95; vol. 66, pt. 3, 1933, 119–21; vol. 68, pt. 2, 1935, 134; *Report of the Minister of Lands and Forests of the Province of Quebec* (Quebec: Redempti Paradis, 1940), 150.

41 Tremblay, *La Tragédie du lac St-Jean.*

42 Concern over and study of the events of the 'tragedy' is evident in correspondence found in RN, Rivière Saguenay (Grande Décharge); on the 1928 flood, see especially the explanation of O. Lefebvre (chief engineer of the Quebec Streams Commission) in Memo Pour l'Honorable Premier Ministre Re. Inondations au lac St-Jean 1928, copy for the Minister of Lands and Forests Honoré Mercier; "Force majeure" is used by Minister Mercier to J. Ed. Boily of Roberval, 16 novembre 1928. Latreille, "Notes re. lac St-Jean," 19 novembre 1926, and "Mémoire Personnel re. Contrôle du lac St-Jean," 10 octobre 1928, file Rivière Saguenay (Grande-Décharge). Latreille also suggested completing the excavation of the outflow channel in the Grande Décharge as a means of preventing another flood.

43 Amos joined the Streams Commission is late 1914. Additional evidence linking the flooding at Lac St-Jean to the Streams Commission's study of the Peribonka is found in O. Graham (QSC assistant chief engineer) to A.B. Normandin (acting chief of the Hydraulic Service), 26 février 1940, file Lac Manouan, RN. Graham explains that the Commission's study was carried out in response to the request of Lands and Forests in part "remédier aux inondations du lac St-Jean, telles que celles de 1928, etc."

44 *Dix-Septième Rapport de la Commission des Eaux Courantes de Québec*, 1928 (Québec: Rédempti Paradis, 1929), 95–6; *Dix-Huitième Rapport*, 1929, 14, 71–2; *Dix-Neuvième Rapport*, 1930, 15, 146–8; *Vingtième Rapport*, 1931, 14–15, 152–6.

45 Massell, "As though there was no boundary," 202.

46 LAC, RG 14, 1997–98/628, House of Commons, War Expenditures Committee, testimony of R.E. Powell, vol. 19, EE2; R.E. Powell, speech to the Canadian Club of Montreal, 6 February 1956, copy in the BANQQ,

E16, 1960-01-035/90, file 2658; Duncan Campbell, *Global Mission*, vol. 1, chap. 7.

47 These conclusions echo those of Bellavance, "L'État, la 'houille blanche' et le grand capital"; and Massell, *Amassing Power*.

48 Girard and Perron, *Histoire*, 317.

CHAPTER 2

1 Massell, "Power and the Peribonka, a Prehistory: 1900–1930s," *Quebec Studies*, vol. 38, Fall 2004/Winter 2005, 87–103.

2 Entrevue realisé pour le Musée amérindien, Louis-Georges Boivin, 1988, MAM.

3 As Joseph described: "il s'est cassé une veine."

4 Entrevue de périodisation, 1981, Informateur 74/Joseph Boivin, 120 pages, re. 1923 voyage, by Joanne Philippe and Alain Nepton, transcribed by Juliette Bégin; also for this voyage, Fiche d'entrevue de l'informateur no. 74 (Joseph Boivin); and Fiches de description du site de campement for the voyage of 1923; MAM.

5 "Ils buvaient quasiment la moitié de leur fourrure."

6 Entrevue de périodisation, Fiche d'entrevue, and Fiches de description du site de campement, for Boivin's voyage of 1923.

7 "C'était dur. Un canot qui est chargé au lourd, un canot de 16 pieds presque 1000 livres dedans, c'est dur a environner dans l'eau. On avait pas de moteur dans ce temps la. Dans les portages on faisait 6 à 7 voyages. Un voyage qu'on a passé 14 fois."

8 "On avait rien dans ce temps-là, on avait pas de moteur, pas de mappe, même pas de compas. C'est les Indiens qui nous disaient comment ç'était fait. Si on découvrait des territoires nouveaux, comme à la tête du Péribonka, on marchait pis quand on trouvait des lacs, on leur donnait des noms. On savait que ça coulait soit au Bersimis, ou au Lac St-Jean ou au Manicouagan parce qu'on allait chasser jusque sur les hauteurs des terres."

9 "Je suis apercu que ça avait été tout couru puis ç'avait été chassé dans l'année d'avant. Les Boivin, ils ne m'avaient pas envoyés dans le bon terrain."

10 "Je ne veux pas vous retarder, si j'ai à vivre je vais vivre et si j'ai à mourir alors je vais mourir."

11 Fiche d'entrevue de l'informateur no. 13 (Jean-Baptiste Dominique) and Fiches de description du site de campement for the voyage of 1929,

MAM; and Dominique's Memories at www.surlestracesilnu.ca. Accessed July 2007.

12 Conversation with Gérard Siméon, June 2006.

13 Fiche d'entrevue de l'informateur no. 159 (Gérard Siméon) and Fiches de description du site de campement for the voyage of 1931, MAM.

14 Fiche d'entrevue de l'informateur no. 97 (Elie Connolly) and Fiches de description du site de campement for the voyage of 1935, MAM.

15 Fiches d'entrevue de l'informateur 38 (Paul Benjamin) for the voyage of 1936–37, CIP, interview conducted by Jean-Marie Vollant and Margot Bacon-Vachon, 1981.

16 "Chevauchement des terrains de trappe entre Pointe-Bleue et Betsiamites," Conseil de bande de Betsiamites, Bureau politique, 1998, Entrevue avec Monsieur Paul Benjamin for the period of 1937–38.

17 Fiches d'entrevue de l'informateur 38, CIP. For the meaning of Innu terms, I am relying on Géo. Lemoine, *Dictionnaire Français-Montagnais avec un vocabulaire Montagnais-Anglais, une courte liste de noms géographiques et une grammaire Montagnaise* (Boston: W.B. Cabot and P. Cabot, 1901); "Vocabulaire de la chasse," in Frenette, *Betsiamites*, Annex C; and the occasional explanation/translation in the hunters' interviews themselves. Regarding 'Pessamit,' Frenette notes (p. 34) that the name derives from periodic invasions of the river by parasitic lamprey, which prey on salmon.

18 Conversation with Adélard Bellefleur, age 83, at the Pessamit Reserve, 20 May 2008, with translation (Innu to French) by Jack Picard. The date given here, 1938–39, is an approximation, as I merely asked Bellefleur to recount his family's migration "on the eve of World War Two."

19 "Nous savions que nous allions avoir de la misère dans le bois, pourtant nous étions tellement heureux quand nous partions vers le territoire de chasse."

20 "On roulait sur la rivière comme une automobile sur la route, ça roulait vite. On a manqué de nourriture, c'était dur." Fiches d'entrevue de l'informateur 148 (Charles-Henry Picard), 1940, CIP, interview conducted by Rolande Rock in 1981.

21 Julius Lips, *Naskapi Law (Lake St. John and Lake Mistassini Bands): Law and Order in a Hunting Society*, in *Transactions of the American Philosophical Society*, vol. 37, pt. 4 (December 1947): 379–492/451, commercial orders of 1935.

22 For example, Fiche d'entrevue de l'informateur 165 (André Hervieux) for the voyage of 1932–33, CIP, interview conducted by Mathieu Paul in 1981.

23 Arthur Ray, *The Canadian Fur Trade in the Industrial Age* (Toronto: University of Toronto Press, 1990).

24 Lips, *Naskapi Law*, 446.

25 These generalizations are nuanced by D. Brassard in *Occupation et utilisation*, chap. 3. To the northeast of the Saguenay, in this same era, the Cree of James Bay were similarly "separating their religious life into the spheres of bush and town," according to Toby Morantz's excellent study, *The White Man's Gonna Getcha: The Colonial Challenge to the Crees of Quebec* (Montreal & Kingston: McGill-Queen's University Press, 2002), 96.

26 Julius E. Lips, "Notes on Montagnais-Naskapi Economy (Lake St. John and Lake Mistassini Bands)," *Ethnos* 12, 1–2 (January–June 1947): 1–78; and Lips, *Naskapi Law.*

27 J. Allan Burgesse, "Property Concepts of the Lac-St-Jean Montagnais," *Primitive Man* 43 (1945): 1–25.

28 "C'est pas le jardinage qui les interesse. C'est pas l'ouvrage d'homme. Ils sonts des chasseurs" – from the film *Les Montagnais,* based on footage recorded between 1936 and 1940. A number of Provencher's photographs are published in *Histoire et culture innues de Betsiamites*; the originals are found at the BANQQ, Fonds Ministère de la Culture et des Communications (E6).

29 Brassard, *Occupation et utilization*, 37–8.

30 J. Frenette, *Betsiamites*, 145: "Le cycle de vie des Montagnais de Betsiamites ... est essentiellement orienté vers l'exploitation des ressources naturelles. En cela, il respect sans doute le mode de vie habituel des chasseurs du subarctique."

31 D. Brassard, *Occupation et utilisation*, 81; and Entrevue de périodisation, Informateur 16 (Jack Germain), conducted by Alcide Blacksmith, 1981, 7, MAM. We might well ask, as a means of measuring the vitality of Indian life, what portion of the Lac St-Jean band lived year-round on the reserve in the 1930s. The geographer Raoul Blanchard visited the Point Blue Reserve in the autumn of 1932 and noted that roughly half the families of the community (being Indians) made trapping and hunting their occupation, which would be a decline from the two-thirds that did so in/around 1910. (*L'Est du Canada Français*, Tome 2, Paris: Librairie Masson, 1935, 133). However, it is not clear that Blanchard's figures are to be trusted (see Veilleux, "Histoire," 154, note 56). Nor is it clear that such rapid demographic change would have been due to hunters settling down rather than a relative population increase among the sedentary mixed-bloods. Ethnologist Lips suggests as much in *Naskapi Law* (p. 397); and anthropologist Burgesse dismissed the villagers altogether from any

assessment of Native life. They "are really half-breeds who have been imported to work the farm lands of the reserve and though they comprise a political party of some importance, have little or nothing of the cultural background of the hunting Indians. For all practical purposes they may be regarded as whites" ("Property Concepts," pp. 1–2).

32   Lips, *Naskapi Law*, 399.

33   See, for example, Donald Wallace, *Market Control in the Aluminum Industry* (Cambridge: Harvard University Press, 1937); Charles Carr, *Alcoa: An American Enterprise* (New York: Rinehart and Company, Inc., 1952); Paul Clark, "Rivers of Aluminum: The Story of Alcan" (1978 manuscript held by Alcan in Montreal); or, most recently, José Igartua, "'Corporate' Strategy and Locational Decision-Making: The Duke-Price Alcoa Merger, 1925," *Journal of Canadian Studies* 20 (1985–86): 82–101.

34   On Duke, see D. Massell, *Amassing Power*.

35   Congressman De Lacy, Hearings Before the Special Committee to Study Problems of American Small Business, United States Senate, Part 58: The Shipshaw Contract: II, 11 April 1945, 7038. On the Roosevelt administration's hostility toward the international aluminum cartel generally, see, for example, Louis Marlio, *The Aluminum Cartel* (Washington: The Brookings Institution, 1947), chap. 1.

36   Paul Clark, quoted by Litvak and Maule, *Alcan Aluminium Limited*, 40ff.

37   Ibid.

38   Biographical information gleaned from Campbell, *Global Mission*, vol. 1; the company newspaper *The Aluminum Ingot*, which began publication in October 1942 (copies available at Alcan, Montreal); and Jack Hight, "Kingdom of the Saguenay; Canada's Sprawling Aluminum Giant," *The Iron Age*, 5 April 1945, copy at Alcan, Montreal.

39   Canadian Aluminum Exports (in pounds), Exhibit 5 of House of Commons, War Expenditures Committee, LAC, RG 14. Pre-war exports peaked at 77 million pounds to Britain (in 1939), 42 million for Japan (also in 1939), 25 million for the US (in 1937), and 12 million to Germany (in 1938).

40   Campbell, *Global Mission*, vol. 1, 144, 247.

41   Massell, *Amassing Power*, chaps. 4–5.

42   Clark, "Rivers of Aluminum", 226.

43   F.H. Cothran, "Pipmaukan Storage Development, Preliminary Study," 15 February 1939, carried out for Price Brothers, and found in the Archives d'Hydro-Québec, Fonds Régie Provinciale de l'électricité (P3), file 173. Cothran had been an engineer with J.B. Duke's Southern Power Company and served under William States Lee, first to design the

Saguenay's hydroelectric scheme and then, in the 1920s, as construction foreman of the Duke-Price Power Company's Isle Maligne power plant.

44   Clark, "Rivers of Aluminum"; and Campbell, *Global Mission*, vol. I, 267.

45   F.L. Lawton, "The Manouan and Passe Dangereuse Water Storage Developments."

46   Dubose to Normandin, 31 December 1937, RN, file Lac Manouan.

47   Dubose to Normandin, 13 November 1937; and Dubose to Bédard, 3 December 1937; both RN, file Lac Manouan.

48   Dubose to Normandin, 27 January 1938, RN, Lac Manouan.

49   Dubose to Normandin, 28 October 1938, RN, Lac Manouan.

50   Petition to accompany a letter dated 17 May 1939 to the Honourable Maurice Duplessis from Aluminum Company of Canada, Ltd., RN, Lac Manouan. On Quebec hydro policy, see Bellavance, "L'État" and Massell, *Amassing Power.*

51   Normandin to Duplessis, 29 mai 1939, RN, Lac Manouan.

52   Normandin was hired by Arthur Amos, a Liberal, in the early 1920s and served under the original Hydraulic Service chief until Amos was sacked by Duplessis in 1936–7.

53   Normandin to Côté, 21 novembre 1939, RN, Lac Manouan.

54   T.D. Bouchard, *Mémoires* (Montréal: Éditions Beauchemin, 1960), 108; and Jean-Guy Genest, "L'Élection provinciale québécoise de 1939" (Thèse (histoire), Université Laval, 1968), 16.

55   Herbert Quinn, *The Union Nationale: A Study in Quebec Nationalism* (Toronto: University of Toronto Press, 1963), chap. 5; Robert Rumilly, *Maurice Duplessis et son temps*, Tome I (1890–1944) (Montreal: Fides, 1973); Leslie Roberts, *The Chief: A Political Biography of Maurice Duplessis* (Toronto and Vancouver: Clarke, Irwin and Co. Ltd., 1963); Gérard Boismenu, *Le Duplessisme: Politique économique et rapports de force, 1944–1960* (Montréal: Les Presses de l'Université de Montréal, 1981); and Michael Oliver, *The Passionate Debate: The Social and Political Ideas of Quebec Nationalism, 1920–1945* (Montreal: Véhicule Press, 1991).

56   Jean-Guy Genest, "L'Élection provinciale."

57   Antonio Barrette, *Mémoires* (Montreal: Librairie Beauchemin, Limitée, 1966), 50.

58   For example, séance de 10 mars 1943, *Débats de l'Assemblée Législative*, draft version not yet published.

59   LAC, RG 14, House of Commons War Expenditures Hearing, testimony of R.E. Powell, vol. 20, E-5, CC-3; and Albert Whitaker, *Aluminum Trail* (Montreal: Alcan Press, 1974), 214–15.

60 Campbell, *Global Mission*, vol. 1, 252–6; and Powell's testimony in the War Expenditures Hearings, vols. 20–21.

61 Powell notes the connection between British aluminum contracts and Lac Manouan's creation in House of Commons War Expenditures Hearings, vol. 21, B-7. Various other Alcan executives/employees confirm the link, including Lawton, "The Manouan … Developments"; N.S. Crerar, "History of the development of the Saguenay Power System" (1958 address); Whitaker, *Aluminum Trail*; and Paul Clark, "Rivers of Aluminum": these papers, addresses, and memoirs available at Alcan Aluminium Limited in Montreal.

62 On Godbout, see Jean-Guy Genest, *Godbout* (Sillery: Les éditions du Septentrion, 1996), which is based on Genest's thesis of 1977.

63 *Débats de l'Assemblée Législative*, draft version, séance de 15 mai 1941.

64 Geoffrion to Côté, 10 novembre 1939, RN, Lac Manouan. Geoffrion was a director of Aluminium Limited from the company's inception in 1928 until his death in 1946; see Campbell, *Global Mission*, vol. 1.

65 On Amos and the inception of the Hydraulic Service, see Massell, *Amassing Power.*

66 Normandin's intense concern for provincial revenues is made clear by his comparison of Lac Manouan's potential income with pre-existing reservoirs in Quebec, including those on the Matawin, Lièvre, and Gatineau rivers, RN, Lac Manouan.

67 Normandin to Côté, "Rapport Préliminaire, Rivière Peribonca, (Lac Manouan)", draft of 21 novembre 1939, RN, Lac Manouan. Regarding Lac St-Jean storage, the company paid a power tax on power produced at Isle Maligne in excess of 200,000 horsepower. See Massell, *Amassing Power*, chap. 6.

68 Normandin's "Project," 1 fevrier 1940, RN, Lac Manouan.

69 R.E. Powell to A. Godbout, 27 February 1940 and 6 April 1940, Fonds Adélard Godbout, dossier Aluminium Company of Canada, BANQQ.

70 *Statutes of the Province of Quebec*, 8 George V (1918), chaps. 68–70. In 1925, the body of laws respecting waters and waterpowers were consolidated under the Water-Course Act: see *The Revised Statutes of the Province of Quebec*, 1925, vol. 1, chap. 46. Division VI concerns "The Construction and Maintenance of Reservoirs for the Storage of the Water of Lakes, Ponds, Rivers and Streams."

71 Powell to Godbout, 6 April 1940, Fonds Adélard Godbout, dossier Aluminium Company of Canada, BANQQ.

72 F. L. Lawton, "The Manouan … Storage Developments."

73 "Sommaire du projet de Aluminum Company of Canada Limited, 4 avril 1941, in the file Rivière Péribonca (Passe Dangereuse), RN; and Normandin to Côté, "Rapport Préliminaire, Rivière Peribonca, (Lac Manouan)," draft of 26 février 1940, RN, Lac Manouan.

74 Draft No. 3, sent to Raymond Latreille 20 July 1940, RN, Lac Manouan.

75 *Débats de l'Assemblée Législative,* draft version, séance de 28 mars 1944.

76 Côté to Powell, 1 May 1940, RN, Lac Manouan.

77 Raymond Latreille, "Sommaire de project de Aluminum Company of Canada Limited," 4 avril 1941; and Latreille, "Comparaisons illustrant les concessions accordées dans le passé a 'Aluminum Co. of Canada, Ltd.' ou aux autres companies dont elle a le controle," 8 mai 1941; both RN, Passe Dangereuse. Copies are also found in the Fonds Godbout, BANQQ.

78 The Aluminum Company completed its plans and specifications in July 1940. Various drafts of the Order, as well as correspondence regarding these, are found in RN, Lac Manouan.

79 *Report of the Minister of Lands and Forests of the Province of Quebec for the Year ending March 31, 1941* (Quebec: Redempti Paradis, 1942), Appendix 10: Annual Report of the Hydraulic Service. Normandin stepped down 8 July 1940; Latreille took the title of chief engineer on 26 October.

80 The Order is found as Arrête en conseil 2958, Concernant ... lac Manouan, 8 août 1940, Fonds Conseil exécutif Québec, BANQQ.

81 Lawton gives us the construction schedule in "The Manouan and Passe Dangereuse Water Storage Developments."

82 For example: Clark, "Rivers of Aluminum," 225ff.; Whitaker, *Aluminum Trail,* 216; and Campbell, *Global Mission,* vol. 1, 265ff. Another colourful rendering is by Hugh Jeffries, "Up north, there's mystery," *Liberty,* 26 April 1941, copy found in RN, Lac Manouan.

CHAPTER 3

1 See also Massell, "As though there was no boundary."

2 Clark, "Rivers of Aluminum," 230.

3 On the timing of technical preparation for dam building, I'm drawing on Clark, "Rivers of Aluminum," 231; Dubose to Côté, 7 March 1941, RN, Passe Dangereuse; and "Passe Dangereuse Power Project – Preliminary Plan Showing Structures and Lands Affected," 8 January 1941, in file P 4121-02-41-5546, Aluminum of Canada Ltd., found in the Centre d'expertise hydrique du Québec.

4 "Notes au sujet de l'entrevue au bureau de Premier Ministre la semaine dernière avec M. Powell de Aluminum Corporation, 27 janvier 1941," RN, Passe Dangereuse.

5 Dubose's "Memorandum" of 12 September 1942, BANQQ Fonds Adélard Godbout, Aluminum Company of Canada.

6 Dubose to Latreille, 31 January 1941, and Latreille to Dubose, n.d., RN, Passe Dangereuse.

7 Dubose to Latreille, 31 March 1941; and Latreille to Dubose, 2 April 1941; both RN, Passe Dangereuse.

8 Dubose to Côté, 7 March 1941; and Dubose, "Memorandum Re. Proposed Construction Programme at Passe Dangereuse," 22 March 1941; both RN, Passe Dangereuse.

9 Lac Manouan could hold 56 billion cubic feet of water.

10 Memorandum of A.-Euclide Paré, "Rivière Péribonca," 12 mars 1941; memorandum of Chs.-Ed. Deslauriers, "Rivière Péribonca," 25 mars 1941; memorandum "Rivière Péribonca," 27 mars 1941, and "Rivière Péribonca," 7 avril 1941; all RN, Passe Dangereuse. The amounts varied according to whether Aluminum intended the waterpower development only, with limited storage, or the storage development only, or a combination of the two.

11 Massell, *Amassing Power*, chap. 1.

12 Latrcille, "Sommaire du projet de Aluminum Company of Canada Limited," 4 avril 1941, RN, Passe Dangereuse.

13 Latreille's memoranda of 4 April and 8 May 1941 are found in BANQQ Fonds Godbout, dossier Aluminum Company of Canada, both documents submitted to the premier by his minister of lands and forests.

14 Geoffrion to Côté, 7 avril 1941; Côté to Dubose, 8 April 1941; both RN, Passe Dangereuse.

15 Côté to Godbout, 4 avril 1941, BANQQ Fonds Godbout, dossier Aluminum Company of Canada.

16 Geoffrion to Côté, 7 avril 1941; Dubose, "Memorandum," 25 April 1941; Côté to Dubose, 28 April 1941; Dubose to Côté, 29 April 1941; all RN, Passe Dangereuse.

17 Massell, "As though there was no boundary."

18 Latreille, "Commentaires sur le mémoire de M. Dubose," RN, Passe Dangereuse.

19 Latreille to Côté, 12 mai 1941, BANQQ Fonds Godbout, dossier Aluminum Company of Canada.

20 This aspect of the 1922 agreement is documented in Massell, *Amassing Power*, chap. 6.

21   Latreille, "Comparaisons illustrant les concessions accordées dans le passé a 'Aluminum Co. of Canada, Ltd.' ou aux autres companies dont elle a le controle," 8 mai 1941, RN, Passe Dangereuse, and also the Fonds Godbout, dossier Aluminum Company of Canada.

22   O. Lefebvre to P.-É. Côté, 13 mai 1941; and reply, 17 mai 1941; both RN, Passe Dangereuse.

23   Latreille to David Clerk of Quebec's Office du Drainage, 7 octobre 1941 and reply 16 octobre, RN, Passe Dangereuse.

24   See letters of Onésime Tremblay found in *Le Devoir*, 7 octobre 1943, "A propos de la 'tragédie' du Lac-Saint-Jean," and/or "Sur une déclaration de M. Wilfrid Hamel," *Le Devoir*, 6 juillet 1943, clippings in RN, Passe Dangereuse.

25   L.A. Bossé to Sous-Ministre Bédard, 22 mai 1941; Latreille to Dubose, 26 May 1941; Latreille to A.-O. Dufresne, 26 mai 1941; Bédard to Bossé, 26 mai 1941; Dubose to Latreille, 30 May 1941; Bossé to Bédard, 25 juin 1941, and reply 2 juillet; J.X. Mercier to Latreille, 19 juin 1941 and reply 20 juin 1941; "Terrain fermé à la prospection du Lac-St-Jean," *L'Evènement*, 23 juin 1941; and Order–in-Council No. 1509, 13 June 1941; all RN, Passe Dangereuse.

26   Aluminum backed off its original plan to construct the full 120-foot dam immediately because of financial constraints. Instead, the Company planned to construct it in stages: to hold back 70 feet of water initially, and then eventually 110 feet.

27   Girard and Perron, *Histoire*, chaps. 9, 13; Dany Côté, *Histoire de l'industrie forestière*. Quebec Pulp and Paper had operated at a loss since the Depression years, and owed $1.8 million to the Quebec Streams Commission for water rental of Lac Kenogami. The company was controlled, in equal share, by Consolidated Paper and Price Brothers. In October 1942, the Quebec Government's minister of lands and forests Pierre-Émile Côté declared the company bankrupt as a means of reopening the firm's Chicoutimi mill and getting the debt to the province repaid.

28   Gustafson, "The Quebec Pulp and Paper Corp. Freehold Estate on the Peribonka and Manouan rivers," Bibliothéque et Archives nationales du Québec, Montreal, Fonds Régie Provinciale de l'électricité (E21).

29   Correspondence of RN, Passe Dangereuse, beginning with Stewart McNichols, president of Quebec Pulp and Paper Corporation, to McNeely Dubose, 17 July 1941. Aluminum and the timber/paper executives had begun their negotiations in May 1941.

30 Prud'homme to Sénateur Donat Raymond, 13 avril 1942, BANQQ Fonds Godbout, Aluminum Company of Canada. See also Prud'homme to Côté, letters of 7 and 8 avril 1942.

31 Bédard to Côté, 14 avril 1942, BANQQ Fonds Godbout, Aluminum Company of Canada.

32 Côté to Godbout, 20 avril 1942, BANQQ Fonds Godbout, Aluminum Company of Canada.

33 Order-in-Council 893, 3 April 1943, Fonds Conseil exécutif, BANQQ.

34 Bird to Francois Faure, vice-president, Consolidated Paper Corporation Limited, 27 November 1941, RN, Passe Dangereuse.

35 Dubose to Bédard, 19 July 1941, RN, Passe Dangereuse.

36 Order-in-Council 893, 3 April 1943, Fonds Conseil exécutif, BANQQ.

37 A. Bédard to Bird, BANQQ, E21 (Lands and Forests), 1960-01-038/11, file Aluminum Company.

38 The three-member committee on forestry interests performed its work and fixed damages by the spring of 1944, one full year after the Passe Dangereuse dam was completed.

39 McNichols to Côté, 23 October 1941, RN, Passe Dangereuse.

40 *Statutes of the Province of Quebec 1941*, art. 34, chap. 154.

41 Matthew Evenden, *Fish Versus Power: An Environmental History of the Fraser River* (Cambridge: Cambridge University Press, 2004), 100–2.

42 Latreille to Charles Frémont, surintendant de la Chasse et de la Pêche, 11 septembre 1941, RN, Passe Dangereuse.

43 Richard to Latreille, 25 septembre 1941, RN, Passe Dangereuse.

44 Dubose to Latreille, 27 September 1941, RN, Passe Dangereuse.

45 Richard to Latreille, 30 septembre 1941, RN, Passe Dangereuse.

46 Percy Nobbs, *Salmon Tactics* (Boston and New York: Houghton Mifflin Company, 1935), vii, 135.

47 Percy Nobbs, "Anticipated Effects of the Passe Dangereuse Dam," 25 February 1942, found in file 00014 22, Alcan, Montreal; and Percy Nobbs, "Anticipated Effects of the Passe Dangereuse Dam," 26 October 1942, RN, Passe Dangereuse. Dr Vladykov also submitted a report dated 14 December 1942, but we have not found a copy.

48 Omer Bernier, "Report on Ouananiche Fishing Conditions in Lake St. John" for Saguenay Power Company, Ltd., 29 March 1940, and subsequent correspondence, of file 00014 22, Alcan, Montreal.

49 Dubose to Latreille, 3 February 1943, RN, Passe Dangereuse.

50 Latreille to Hamel, 25 février 1943, and Bird to Hamel, 16 March 1943, RN, Passe Dangereuse.

51 Latreille to Bird, 10 mai 1943, RN, Passe Dangereuse.

52 Latreille to Hamel, 7 avril 1943, RN, Passe Dangereuse.

53 Order-in-Council 893, 3 April 1943, and 911, 6 April 1943; both Fonds Conseil exécutif, BANQQ.

54 "La Compagnie d'Aluminum a payé $50,000 à titre de compensation," *L'Evènement*, 8 mai 1943, and a similar article in *L'Action Catholique*, 8 mai 1943; clippings in RN, Passe Dangereuse.

55 Donald Worster describes the transition from utilitarian "conservation" to preservationist "ecology" in *Nature's Economy: A History of Ecological Ideas* (Cambridge: Cambridge University Press, 2nd ed., 1994), pt. five.

56 Girard and Perron, *Histoire*.

57 Evenden, *Fish Versus Power*.

58 Latreille to Richard, 27 octobre 1942, BANQQ, Fonds Richesses naturelles, 1996-06-001/21, Possibilité d'expansion des centrales hydroélectriques pour fins de guerre.

59 For example, Dubose to Latreille, 21 October 1941, RN, Passe Dangereuse.

60 Correspondence re. the tax issue begins in RN, Passe Dangereuse, with V. Edward Bird to R. Latreille, 8 October 1941. Particularly useful are two memoranda which summarize these negotiations: Latreille's "Mémoire à l'honorable P.-Emile Côté," 17 décembre 1941, RN, Passe Dangereuse; and Dubose's "Memorandum" of 12 September 1942, BANQQ Fonds Godbout, Aluminum Company of Canada.

61 Dubose to Latreille, 3 January 1942, and reply 7 January; both RN, Passe Dangereuse. On Dubose, see Campbell, *Global Mission*, vol. 1.

62 Latreille to Lefebvre, 18 décembre 1941, RN, Passe Dangereuse.

63 "Réservoir Passe Dangereuse," 10 février 1943, likely written by Latreille, RN, Passe Dangereuse; and Powell to Godbout, 3 February 1943, BANQQ Fonds Godbout, Aluminum Company of Canada.

64 Latreille to Dubose, 26 February 1943; and Bird to Latreille, 26 February, 1943, RN, Passe Dangereuse.

65 Latreille to Dubose, 4 September 1943, BANQQ Fonds Godbout, Aluminum Company of Canada.

66 Yvon De Guise, "Mémoire," 27 juillet 1944, RN, Passe Dangereuse.

67 Correspondence in this matter begins with Bird to Bédard, 4 June 1945, RN, Passe Dangereuse.

68 Arrête en conseil 893, 2 avril 1943, Fonds Conseil exécutif, BANQQ.

69 *The Revised Statutes of the Province of Quebec 1925*, vol. 1 (Quebec: Ls. A. Proulx, 1925), 807.

70 On dam construction, see Lawton, "The Manouan and Passe Dangereuse Water Storage Developments"; and the various corporate memoirs

including Clark, "Rivers of Aluminum," 230–4. With the dam built, Aluminum moved to pay flood damages to private interests, and the provincial government verified the size of the flooded zone via aerial and canoe survey of the reservoir. Once Quebec Power declared bankruptcy, Quebec bought up bits of its land for cession to Aluminum and passed an Order-in-Council permitting this process: in short the endgame was performed and loose ends were tied.

71   Massell, *Amassing Power*, chaps. 6–7.

72   Geoffrion to Pierre-Émile Côté, 24 janvier 1941, and Olivier Lefebvre to Pierre-Émile Côté, 29 janvier 1941, BANQQ Fonds Godbout, Hydro-électricité. Wrote Geoffrion: "La loi, telle qu'elle se lit, ne permet l'approbation, soit par le lieutenant-gouverneur en conseil, soit par la Régie, qu'avant la construction ... Je suggérerais donc d'ajouter dans le Chapitre 46 S.R.Q. un troisième paragraphe à l'article 5 [re. approval of a dam or dike] qui lirait ainsi: 'Néanmoins, le lieutenant-gouverneur en conseil peut, après que ces ouvrages ont été construits en tout ou en partie, ou meme exploités, en approuver l'emplacement et les plans et devis lorsque telle approbation, quoique requise par la loi, n'aura pas été obtenue au préalable.' Je suggérais le même amendement à l'article 57 [regarding the construction of reservoirs] afin de couvrir non seulement le cas des développements mais aussi le cas des emmagasinements." The amendment was one of two sought; the other dealt with corporations' ability to expropriate land for electrical transmission lines.

73   Latreille to Lefebvre, 26 août 1941, RN, Passe Dangereuse.

74   Bird to Côté, 23 August 1941, RN, Passe Dangereuse.

75   "'Aluminum of Canada' veut affermer les eaux du bassin du Saguenay," *L'Evènement*, 26 Septembre 1941.

CHAPTER 4

1   *Industrial Mobilization for War: History of the War Production Board and Predecessor Agencies, 1940–1945*, vol. 1 (New York: Greenwood Press, 1969, originally 1947), iii, 988–92 for a list of studies.

2   As its title suggests, Philip J. Funigiello, *Toward a National Power Policy: The New Deal and the Electric Utility Industry, 1933–1941* (Pittsburgh: University of Pittsburgh Press, 1973) encroaches on the wartime period but focuses on the evolution of federal policy, rather than electrical power as an ingredient of war production. Even less is found in the Twentieth Century Fund's *Electric Power and Government Policy: A Survey of the Relations Between*

the Government and Electric Power Industry (New York: The Twentieth Century Fund, 1948), 732–9.

3 Julius Krug in press release of 21 July 1941, found in Marks Papers, file Power – Aluminum; Harry Rohs, "Utilities Set to Meet All Power Demands Now and in the Future," *Wall Street Journal*, April 1941, Olds Papers, file Clippings: Power for War.

4 Funigiello, *Toward a National Power Policy*, 230, chap. 9.

5 Olds Papers, files Minutes of Meetings; and Clippings, 1939–1940.

6 "Capacity," 13 January 1940, 96–7; "Power Still in Politics," 29 June 1940, 2054; "Power Supply Adequate to Meet Country Demand," 8 June 1940, 113, 9; *Electrical World* 113 (1940). "Power for Defense – a Job for the Electric Utilities," 19 October 1940, 1150; *Electrical World* 114 (1940).

7 In all, the US government installed some 2.5 million kilowatts of power capacity by 1943, some four-fifths of this being hydroelectricity. See Problems of American Small Business Enterprises, United States Senate (Washington: US Government Printing Office, 1945) [hereafter Problems of American Small Business] pt. 57, 6880–1 graphs. The Pacific Northwest and TVA systems together comprised roughly 1.7 million kilowatts of the total.

8 Testimony of Leland Olds, the Truman Committee, pt. 3, 868.

9 J. Carlyle Sitterson, *Aircraft Production Policies Under the National Defense Advisory Commission and Office of Production Management, May 1940 to December 1941* (Civilian Production Administration, Historical Reports on War Administration, Special Study No. 21, 1946), 7.

10 Funigiello, *Toward a National Power Policy*, 251.

11 The formal title was Heat, Light and Power Unit.

12 The Advisory Commission (or NDAC) became the Office of Production Management (OPM) in January 1941, which, in turn, gave way to the WPB.

13 This would hardly be the last time that public- and private-power advocates clashed within the US war bureaucracy, or that Olds sought to defend the FPC's power turf from wartime bureaucratic encroachment.

14 Stettinius was US Steel's 39-year old chairman of the board and NDAC's advisor on industrial materials. Stettinius, in turn, was advised on power matters by Gano Dunn, of J.G. White Engineering Corporation, who had formerly fought against the TVA but now found himself working on government's behalf. Stettinius was advised on mining and mineral matters by William Batt, formerly of SKF Industries. Julius Krug, TVA's manager of power, would take over a newly constituted Power Branch of the OPM in the spring of 1941 as the war bureaucracy continued to

evolve; and Krug, like Kellogg before him, would experience significant conflict with Leland Olds.

Individuals can easily be identified in *Industrial Mobilization for War.* The Kellogg-Olds relationship can be discerned in War Production Board records, 016.868R, Power Branch, OPM, Weekly Progress Reports; or in the Olds Papers, for example, Thomas Tate's memo to Olds of 19 September 1939, file Editorial, *Electrical World.* On the Olds-Krug conflict, see Donald M. Nelson, *Arsenal of Democracy: The Story of American War Production* (New York: Harcourt, Brace and Company, 1946), 365–6.

15 Hearings Before the Subcommittee of the Committee on Appropriations United States Senate on S.J. Res. 285, 10 July 1940, 15, 19.

16 Marguerite Owen, *The Tennessee Valley Authority* (New York: Praeger Publishers, 1973). Chap. 5; Eliot Janeway, *The Struggle for Survival: A Chronicle of Economic Mobilization in World War II* (New Haven: Yale University Press, 1951), 235; brochure "Federal Columbia River Power System," 17 pages, ca. 2001.

17 Alcoa later expanded that powerhouse to generate 60,000 horsepower by installing additional turbine-generators; by the Second World War, the company was obtaining some 100,000 horsepower from this same plant. (C.W. Kellogg, weekly report, 4 April 1941, WPB 016.868R.)

18 Carr, *Alcoa*, and R.T. Danforth, *Aluminum Company of America and Subsidiaries, Massena, New York, Chronological Notes*, copy at Alcoa Primary Metals, North-East Smelting Region, Massena Operations.

19 On Cedar Rapids, see André Bolduc, Clarence Hogue, and Daniel Larouche, *Québec: un siècle d'électricité*, 73–7; and Jean Louis Fleury, *Une Ligne et des hommes* (Montreal: Libre Expression, 1991).

20 Only in the late 1950s, with the construction of the St Lawrence Seaway, including the enormous Moses-Saunders Power Dam, would the aluminum company, along with the region as a whole, find an alternative power source for large-scale industrial development.

21 The other concern was Alcoa's need for 30,000 kilowatts of power at the Badin plant in the Carolinas pending the completion of the TVA's Santee-Cooper project.

22 Kellogg, quarterly report, 5 December 1940, National Archives and Records Administration of the United States, WPB, file 013.3112; notes of conversation between Olds and Hickerson (Dept. of State), 20 May 1941, Olds Papers, file Canada; testimony of William Batt, Problems of American Small Business, pt. 57, 6865.

23 Kellogg, weekly report, 19 November 1940, WPB, file 016.868R; Kellogg, weekly reports of 14 January 1941, 21 February 1941, WPB, file

016.868R; Olds Papers, file Alcoa Power Supply, Olds to Berle, 24 June 1941.

24 Handwritten note, Olds Papers, file Notes (L.O.): Gano Dunn (this was likely the same conversation, dated 31 October, noted by Olds in memo for Senator James Murray, 23 June 1945, Olds Papers, file Aluminum); Kellogg, weekly reports of 4 February 1941, 10 March 1941, 25 November 1940, WPB, file 016.868R. G.R. Holden to Harriman, 18 February 1941, WPB, file 523.1; Kellogg, weekly report, 10 March 1941, WPB, file 016.868R. Alcoa vice-president I.W. Wilson also pressed Olds on the Massena matter. Olds told Wilson that any participation by Alcoa itself in the power talks would only delay matters.

25 G.R. Holden to Moffett, 10 February 1941, WPB, file 523.2; G.R. Holden to W.A. Harriman, 18 February 1941, WPB, file 523.1; Holden to W. Batt, 3 May 1941, WPB, file 523.433; Meeting of Sub-Committee on Aluminum and Magnesium, 3 May 1941, WPB, file 523.3. On the Alcan-Alcoa contract, see Campbell, *Global Mission*, 247.

26 On Olds's long-standing involvement with the Seaway, see Olds Papers, Box 58, files marked Historical Data; Agreement with Canada, 1941; Canada; State Department; and President's Messages. On Roosevelt's long-standing "dream" to make a comprehensive development of the Seaway, see, for example, Adolf Berle's memo of conversation with Olds and FDR, 11 September 1940, in *Foreign Relations of the United States* 1940, vol. III (Washington: Department of State), 150–1.

27 See, for example, Berle Papers, St Lawrence Seaway (1940). In Berle's address to the Great Lakes Seaway and Power Conference in Detroit, 5 December 1940, he explained that the Seaway was now under survey because "as Olds will tell you ... the power resources of Canada are already taxed to capacity for defense needs; and because already we need more power on the American side of the St Lawrence for urgent defense needs. Specifically, we need to manufacture more aluminum at the plants at Massena." Roosevelt's message of the same date, which Berle read to the gathering, contains much the same message, as does FDR's Special Message to Congress of 17 October 1940 on the St Lawrence Development (Olds Papers, President's Messages). We find Berle and Olds and their Seaway committee colleagues working out political strategy, including "the national defense angle," on 22 November 1940 (Berle Papers, St Lawrence Seaway (1940), re. St Lawrence Waterway); and as early as May 1940, Berle suggested to FDR that the administration "push" the Seaway on the basis of "more electrical power" for "defense" (Berle to Roosevelt, 16 May 1940).

28 We look in vain in the papers of Leland Olds or Adolph Berle, as well
    as in the records of the US Department of State, for documentation
    that either man laboured in good faith to win additional power for the
    Massena plant during the fall and winter of 1940–41. By contrast, their
    efforts to negotiate and then publicly market a St Lawrence treaty as a
    national defence measure is everywhere in evidence. By late 1942, in any
    case, the Seaway was a dead letter in the US Senate. See Carleton Mabee,
    The Seaway Story (New York: The Macmillan Company, 1961), chap. 11.

29 Kellogg, weekly report, 21 April 1941, WPB, file 016.868R.

30 "Start Draft on New St Lawrence Treaty," 28 December 1940, *Electrical
    World* 114, 1946.

31 Kellogg, weekly reports, 22 March 1941, 1 March 1941, and 28 March
    1941, WPB, file 016.868R; further details are revealed in Kellogg's Diary
    Notes on Massena, WPB, file 610.42.

32 Symington's Liberal values and loyalty to Howe are nicely demonstrated
    in LAC, John Wesley Dafoe Papers (MG 30-D45), for example, Symington
    to Dafoe 19 February 1941 and 4 October 1943. On Symington and
    Howe, see Philip Smith, *It Seems Like Only Yesterday: Air Canada, the First
    50 Years* (Toronto: McClelland and Stewart, 1986); also Robert Bothwell
    and William Kilbourn, *C.D. Howe: A Biography* (Toronto: McClelland and
    Stewart Limited, 1980), where "great leader" (p. 354) is the authors'
    phrase and "chief builder" (p. 349) is the *Toronto Star*'s. On Symington
    and wartime control work, see Matthew Evenden's "Lights Out:
    Conserving Electricity for War in the Canadian City, 1939–1945," *Urban
    History Review* 34, 1 (2005): 88–99. Other biographical information on
    Symington derives from *Canadian Who's Who*, vol. 11, 1936–7 (Toronto:
    Murray Printing Co. Ltd., 1936), 1048; *Who's Who in Canada*, 1945–46
    (Toronto: International Press Ltd., 1946), 554; *Who's Who in Canada,
    1947–8* (Toronto: International Press, Ltd., 1948), 1414; *The Canadian
    Who's Who*, vol. 9, 1961–3 (Toronto: Trans-Canada Press, 1962), 1084.

33 23 August 1940, P.C. 4129, LAC, Munitions and Supply, vol. 54, file 1-1-
    98; see also deputy minister G.K. Sheils to R.A.C. Henry, 26 July 1940,
    and Henry's reply of the same date, LAC, Munitions and Supply, vol. 54,
    file 1-1-98.

34 Quebec's Public Service Board/Régie des services publics replaced
    Duplessis's Provincial Electricity Board in 1940 (which, in turn, Duplessis
    had substituted for the Liberal's Electricity Commission of 1935) and
    became the arbitration and supervisory body for public services and
    utilities. In addition to supervising transportation and communication,
    the PSB held "jurisdiction over the production, the transmission,

distribution and sale of electricity in the Province of Quebec and wide powers respecting service, equipment, apparatus, means of protection, extensions of plants and systems, as well as control of rates and capitalization." (*Annuaire Statistique/Statistical Year Book Quebec 1941*, 352). With the outbreak of war, the PSB's John McCammon undertook a survey of Quebec's electrical utilities to determine/locate any unused capacity of energy that might be employed in wartime industries. With the data in hand, McCammon also prepared summaries of its results, which he conveyed to the Water Power Bureau of Ottawa's Department of Mines and Resources. (See BANQQ, Fonds Richesses naturelles (E20), 1996-06-001/21, file Possibilité d'expansion des centrales hydroélectriques pour fins de guerre, correspondance and memoranda of October–November 1939). In July 1940, by a wartime power control act, the PSB assumed additional powers "over the production, transformation, transmission, distribution and supply of motive power to ensure the best possible utilization thereof." (*Premier Rapport de la Régie des Services Publics*, 1 August 1940 to 31 March 1941, Québec: Rédempti Paradis, 1942, 207). We should note that the Régie's original file dealing with wartime power needs (labelled P-37) seems to have been lost to us; and the file of BANQQ utilized here is but an echo of the original. It contains copies of a portion of the Régie's file, sent by the Régie to Raymond Latreille's Service Hydraulique to keep Latreille informed. Decades later, in 1996, Richesses naturelles passed on the file to the BANQQ along with a mass of other waterpower-related material.

35 Testimony of H. J. Symington, LAC, RG 14, House of Commons War Expenditures Committee, vol. 25, B5–B6; R.A.C. Henry to G.K. Sheils, 11 September 1940, and Memorandum Re Operations of the Power Controller, 3 October 1940, of Munitions and Supply, vol. 54, file 1-1-98. A longer 1943 version of the memo is found in vol. 253, file 196-11-14, in which Symington explains that "Consideration was given to the general setup of the Control, having in mind the fact that each Province was very jealous of its power rights and that provincial jurisdiction should not be interfered with any more than was necessary. Accordingly it was decided that the work should be decentralized as much as possible and the provincial organizations in charge of the power situation in each province were consulted and their assistance secured. The result was that nothing of a drastic nature was done without consultation with the electrical authorities and, in each case, their approval secured. The Control used these organizations in the various provinces as its agents and subsequent results have shown that that was a wise manner in which

to handle the Control." An additional example of Symington's sensitivity
to provincial jurisdiction is found in Symington to Powell, 7 November
1940, BANQQ, Fonds Richesses naturelles, Possibilité d'expansion des
centrales hydroélectriques pour fins de guerre. An overview of Munition
and Supply's task of "Power Control" is provided in J. de N. Kennedy,
*History of the Department of Munitions and Supply: Canada in the Second World
War* (Canada: King's Printer, 1950), chap. 11.

36 Testimony of H.J. Symington, LAC, RG 14, House of Commons War
Expenditures Committee, vol. 25, E5–E6.

37 Testimony of Frank Brown, 5 October 1943, LAC, RG 14, House of
Commons War Expenditures Committee, vol. 7, A11–B1, and H.J.
Symington, vol. 25, F2. Engineer R.A.C. Henry of Munitions and Supply
also went along. On the Saguenay's advantages for aluminum production,
see Massell, *Amassing Power*, 179.

38 A. W. Whitaker, *Aluminum Trail*, 214–15; echoed by Campbell, *Global
Mission*, 253–4. R.E. Powell offers additional details of negotiation of the
British contracts in his testimony before the House of Commons War
Expenditures Committee, vols. 20–1.

39 "Because the mighty Saguenay was dammed," a company publicist put it,
"the quickest way, in the whole of North America, to get more power was
to extend the Shipshaw development" ("This is Shipshaw," *The Aluminum
Ingot*, 30 January 1943, 6).

40 Testimony of Frank Brown, House of Commons War Expenditures
Committee, vol. 7, B7–8, and H.J. Symington, vol. 25, F2, FF1, EE1-4, RG
14, LAC. Other power sites were considered, including on the Ottawa, St
Lawrence, and Nelson rivers, but none could be developed in as timely a
fashion as the Saguenay.

41 Jerome G. Kerwin, *Federal Water-Power Legislation* (New York: Columbia
University Press, 1926).

42 Howe to Symington, 28 October 1940, LAC, Munitions and Supply, vol.
54, file 1-1-98.

43 Bothwell and Kilbourn, *C.D. Howe*, Epilogue.

44 Howe to Symington, 28 October 1940, LAC, Munitions and Supply, vol.
54, file 1-1-98. Symington's thoughts on the matter are expressed in
House of Commons War Expenditures Committee, vol. 25, CC1–2, EE1-
4, RG 14, LAC.

45 Testimony of R.E. Powell, House of Commons War Expenditures
Committee, vol. 19, EE2, RG 14, LAC; and Graham, MP, quoting
Dubose, vol. 7, B6. Of its pre-war capacity to produce 45,000 tons/year

of aluminum, Alcan produced one-third (or less) at the bottom of the Depression, thus 15,000 tons/year. (Powell, vol. 22, A3.)

46   Whitaker, *Aluminum Trail,* 214–15. Expansion included additional Arvida potrooms, installation of oversize waterwheels at Chute-à-Caron and a twelfth generator at Isle Maligne, many miles of transmission lines to divert Quebec power to the Saguenay region, and the construction of the timber storage dam on the Rivière Manouan.

47   Davis's remarks to shareholders quoted by T.L. Brock, "Alcan in the Saguenay – the Formative Years,"(Montreal: Aluminium Company of Canada, Ltd., 1971), 46–7. Re. Davis's anxieties about over-supply during the war itself, see 43–4. Blough's comments are found in "Discussion on the Design of the Shipshaw Power Development," *The Engineering Journal* 27, 8 (August 1944): 449; Powell's comments were made in House of Commons War Expenditures Committee, vol. 23, C6, RG 14, LAC.

48   House of Commons War Expenditures Committee, testimony of Justice C.P. McTague, vol. 13, DD2–3, and H.J. Symington, vol. 25, J2, RG 14, LAC; and Exhibit 1, Annual World Production of Aluminum by Countries.

49   House of Commons War Expenditures Committee, testimony of F. Brown, vol. 7, C2, vol. 9, D4, D11, and H.J. Symington, vol. 25, AA9, J1, RG 14, LAC.

50   Campbell, *Global Mission,* vol. 1, chap. 8; and vol. 2. Arvida's aluminum works, following post-war cutbacks in ingot production, were running at full capacity by 1948.

51   As early as October 1940, Symington was in contact with Alcan regarding power shortages in Quebec and the necessary electrical interconnections to permit aluminum production to proceed at Arvida to serve the British contracts.

52   Symington memo of 21 February 1941 quoted by F. Brown, House of Commons War Expenditures Committee, vol. 7, A10–11 (a similar version was sent to R.E. Powell the same day, as quoted by Clark, "Rivers of Aluminum," 224, and used by Campbell, *Global Mission,* 272); Symington, House of Commons War Expenditures Committee, vol. 25, F1–2, BB2, echoed by Brown vol. 7, for example, C5, RG 14, LAC.

53   Clark, "Rivers of Aluminum," 224, also used by Campbell, *Global Mission,* 272.

54   Robert B. Bryce, *Canada and the Cost of World War II: The International Operations of Canada's Department of Finance, 1939–1947,* edited by Matthew J. Bellamy (Montreal & Kingston: McGill-Queen's University Press, 2005), chap. 5; R.D. Cuff and J.L Granatstein, *Ties That Bind:*

Canadian-American Relations in Wartime from the Great War to the Cold War,
2nd ed. (Toronto and Sarasota: Samuel Stevens, Hakkert and Company,
1977), chap. 4: "The Hyde Park Declaration, 1941: Origins and
Significance."

55 Bryce, *Canada and the Cost of World War II.*

56 "Joint statement by the Prime Minister of Canada and the President of
the United States, issued at Hyde Park, N.Y., Sunday evening, April 20,
1941."

57 J.W. Pickersgill, *The Mackenzie King Record,* vol. 1 (Chicago and Toronto:
University of Chicago Press, University of Toronto Press, 1960), 202.

58 Charles Wiltse, *Aluminum Policies of the War Production Board and
Predecessor Agencies May 1940 to November 1945* (Historical Reports on
War Administration: War Production Board, Special Study No. 22),
36ff.; J. Carlyle Sitterson, *Aircraft Production Policies Under the National
Defense Advisory Commission and Office of Production Management May
1940 to December 1941* (Historical Reports on War Administration: War
Production Board, Special Study No. 21), 124ff.; Gerald D. Nash, *World
War II and the West: Reshaping the Economy (Lincoln: University of Nebraska
Press, 1990),* 100; J.W. Pickersgill, *The Mackenzie King Record,* vol. 1, 191.

59 Morgenthau Diaries, 18 April 1941, F.D.R. Library; "Tentative Estimates
of Canada's Available Surplus Capacity for the Production of Munitions
and Other Supplies," 21 April 1941, Morgenthau Diaries; Howe to
Norman Robertson (acting under-secretary of state, Ottawa), 30
April 1941, King Papers, LAC; House of Commons War Expenditures
Committee, testimony of Brown, D3, RG 14, LAC; Minutes of the War
Cabinet Committee, 21 April 1941, RG 2, LAC.

60 Cuff and Granatstein pose the question and make an effort to answer it in
Ties That Bind, 89ff.

61 C.W. Kellogg, weekly reports of 12 May and 15 May 1941, WPB, file
016.868R; Kellogg, Diary Notes on Massena, WPB, file 610.42.

62 Powell's testimony, House of Commons War Expenditures Committee,
vol. 21, C8, RG 14, LAC.

63 Calculations are based on Wiltse's data in *Aluminum Policies of the War
Production Board and Predecessor Agencies,* Appendix D, 347. Appendix
C, 341, breaks down US aluminum production facilities according to
producer: Alcoa, Reynolds Metals, and the US government (Defense
Plant Corporation). Government-produced ingot amounted to some 55
per cent of the wartime total.

64 Powell, House of Commons War Expenditures Committee, vol. 24,
AA2, RG 14, LAC. Said Frank Brown, Munitions and Supply's financial

adviser, "there was only one place the money [for Shipshaw] could come from and that was from the United States government by way of advance payments." (vol. 7, C11). We can glean from Powell's testimony (vols. 19–24) that Alcan's funds for pre-war and wartime construction of both power and ingot-production facilities in Canada through late 1943, which totalled some $200 million, derived mainly from sources outside Canada: from the United States ($98.5 million), the UK ($76 million), and Australia ($2.5 million). The company raised just $15 million in Canada, from the issue and sale of capital stock.

65 Testimony of R.E. Powell, House of Commons War Expenditures Committee, vol. 21, E1.

66 George Moffett to W.L. Clayton, 13 May 1941, National Archives and Records Administration of the United States, Reconstruction Finance Corporation, Metals Reserve Company, Alcan files, pt. 1; I.W. Wilson to William Knudsen, 12 May 1941, found as Exhibit 149, Truman Committee hearings, pt. 8, 2809.

67 LAC, Minutes of the War Committee of the Cabinet, 21 May 1941, 6.

68 The final draft of Symington's letter, dated 31 May 1941 is found in Reconstruction Finance Corporation, Metals Reserve Company, Alcan files, pt. 1. Arrangements leading to this point are also documented here.

69 Problems of American Small Business, pt. 58, 7015.

70 On Massena's power supply, see J.E. Moore to E.J. Earley, 5 December 1944, Marks Papers, file Power-New York-Massena; and testimony of I.W. Wilson before the Truman Committee, pt. 8, 2659–60. On the St Lawrence plant, see Carr, 249. Danforth notes in *Chronological Notes* that Alcoa broke ground on the Plancor Plant on 4 October 1941; the plant first produced aluminum in late May 1942; it was purchased by Alcoa in 1950.

71 House of Commons War Expenditures Committee, testimony of Symington, vol. 25, AA2–6, RG 14, LAC.

72 House of Commons War Expenditures Committee, testimony of Powell, vol. 24, AA2; testimony of McTague, vol. 13, AA1–EE2, RG 14, LAC. The original files of the Department of Finance document this lengthy process in detail. These are found in the LAC, RG 19, vol. 454, file 111-17-7A, "Special Depreciation Under War Measures Act, Aluminum Company of Canada, Ltd – Agreement (Shipshaw Power Project)." Campbell summarizes these events from the company's perspective in *Global Mission*, vol. 1, 375ff.

73 The chronology also makes for a rich case study in Canadian-American wartime relations. See Massell, 'As though there was no boundary.'

74 A.B. Normandin and J.-W. McCammon to P.-É. Côté, 27 octobre 1941,
BANQQ, Fonds Richesses naturelles, 1996-06-001/21, Possibilité
d'expansion des centrales hydroélectrique pour fins de guerre.
An excerpt: "le gouvernement des Etats-Unis a rendu possible le
développement à Shipshaw, sur le Saguenay, de centrales de 500,000 à
600,000 c.v., et ce à la condition souscrite par le Contrôleur Fédéral de
l'Energie que les livraisons seraient maintenues à Massena, New York,
pour assurer la productivité d'usines existantes..."

75 King to Côté, 22 November 1941, BANQQ, Fonds Richesses naturelles,
1996-06-001/21. King's was a reply to Côté's letter of 30 October asking
the prime minister's view on a request by the Maclaren–Quebec Power
Company to export to Ontario an additional block of power (57,500
horsepower) for the duration of the war. King's letter to Côté was a plea
that the Quebec Government permit this export to Ontario (despite
the obvious political sensitivity of the issue in the province) because the
export continued to render practical the larger Canadian–American
arrangement: i.e., MacLaren-Quebec's energy would partially replace
power that the Hydro-Electric Commission of Ontario was already
exporting to Massena in order to fulfull the terms of the Saguenay/
Massena ingot/power bargain. With King's opinion in hand, Godbout's
government approved MacLaren-Quebec's power export.

76 Côté to King, 28 November; and Côté to Godbout, 29 novembre; and
Latreille to McCammon, 28 November, all BANQQ, Fonds Richesses
naturelles, 1996-06-001/21, Possibilité d'expansion des centrales
hydroélectrique pour fins de guerre.

77 Following the death of Ernest Lapointe, King sought to recruit Godbout
as his chief French-Canadian lieutenant: "You are the only one to
succeed Lapointe just as Lapointe was obviously the only one to succeed
Sir Wilfred Laurier in the leadership of Quebec in the wider arena of
the Dominion .... Frankly, my dear Godbout, I am completely at a loss
to know how the party, the government and the country are to be kept
united unless I have someone from Quebec like yourself at my side, and
that immediately." See King to Godbout, Nov. 30, 1941 in Pickersgill, *The
Mackenzie King Record*, Vol. 1, pp. 290–92 or BANQQ Godbout Papers,
Correspondence entre Adélard Godbout et W.L. Mackenzie King.

78 Maurice Riel to Jean Paré, 12 février 1999; Marthe Godbout-Bussières to
Bernard Landry, 30 mars 2003; Olivier Courteaux, "Heard of Godbout?,"
National Post, spring 2000; "Charest comparé à Godbout," *Le Devoir*, 8
avril 2003; and the texts of the various speeches made (including by
Lucien Bouchard, Jean Charest, and Mario Dumont) at the unveiling of

the Godbout monument in Quebec City, 19 October 2000; all in private collection of Marthe-Godbout Bussières.

79 Symington to Powell, 7 November 1940, BANQQ, E20, 1996-06-001/21, file Possibilité d'expansion des centrales hydroélectriques pour fins de guerre.

80 Symington to John Wesley Dafoe, 4 October 1943 and 30 December 1943, MG30-D45, LAC,; and Symington's testimony, House of Commons War Expenditures Committee, vol. 25, CC1–2, EE1–4, RG 14, LAC.

CHAPTER 5

1 Gerald Nash nicely summarizes the Shipshaw controversy in *World War II and the West*, chap. 5: "New Basic Industries for the West: Aluminum."

2 Massell, "As though there was no boundary," 211 and note 105; *Debates*, House of Commons, Dominion of Canada, Session of 1943, vol. IV, 14 June 1943, 3610ff.; testimony of Powell, House of Commons War Expenditures Committee, vol. 24, KK2, RG 14, LAC; House of Commons, Special Committee on War Expenditures, Minutes of Proceedings, Third Report, 26 January 1944.

3 *Débats de l'Assemblée Législative*, séance de 10 mars 1943, draft version.

4 "Un débat sur le développement hydroélectrique de l'Aluminum," *L'Événement-Journal*, 16 juin 1943; *Débats de l'Assemblée Législative*, séance du mardi 15 juin 1943, draft version.

5 *Débats de l'Assemblée Législative*, séance de 7 avril 1943, draft version.

6 *Débats de l'Assemblée Législative*, séance du mardi 15 juin 1943, draft version.

7 "Aucun contrôle sur l'électricité pour la fabrication du papier," *L'Événement-Journal*, 16 juin 1943, 1, 9.

8 "Les deux barrages ont empêché de grosses inondations, au Lac St-Jean," *L'Événement-Journal*, 26 juin 1943, news clipping in RN, Passe Dangereuse.

9 *Débats de l'Assemblée Législative*, séance du mardi 15 juin 1943, draft version; "M. Godbout essaie de justifier l'exploitation de nos ressources naturelles par les étrangers," *Le Devoir*, 16 juin 1943, 7.

10 R. Chaloult, *Mémoires politiques* (Montreal: Éditions du jour, 1969), 107–45.

11 "M René Chaloult expose ses vues sur le programme provincial et les chefs du Bloc populaire canadien," *Le Devoir*, 13 septembre 1943, 7.

12 "La race canadienne-française n'a pas besoin de la tutelle fédérale," *Le Devoir*, 9 août 1943, 4.

13 "L'Aluminum Company édifiée au mépris de nos lois," *Le Devoir*, 4 octobre 1943, 10; "Assemblées de l'Union Nationale," *Le Progrès du Saguenay*, 7 octobre 1943, 10.

14 Hogue, Bolduc, and Larouche, *Québec: un siècle d'électricité*, chap. 7. Côté's request for a report from Latreille is noted in "Raymond Latreille Quitte La Commission Hydroélectrique de Québec," *Hydro-Presse*, 1966, Archives d'Hydro-Québec.

15 "Avis de la Régie des Services publics concernant la Montreal Light, Heat and Power," *Le Devoir*, 8 octobre 1943, 3.

16 All major newspapers of both languages covered the story. For example: "Le gouvernment passera une loi pour étatiser la M.L.H.&P.," *Le Devoir*, 9 Octobre 1943; "Québec étatisera la M.L.H.&P.," *La Presse*, 9 Octobre 1943; "Quebec to Expropriate M.L.H.&P., Godbout Announces," and "M.L.H.&P. Summoned to Show Why Rates Should Not Be Cut," *The Gazette*, 9 October 1943; "See Godbout's Plan As No Idle Threat," "Mr. Godbout Talks Expropriation," and "Montreal Company Under Two-Way Fire," *The Financial Post*, 16 October 1943. The news even made the *New York Times:* "Quebec May Seize Montreal Utility," 9 October 1943, 19.

17 House of Commons War Expenditures Committee, testimony of Symington, vol. 25, EE2–EE4, RG 14, LAC.

18 "Quebec Plans For Own Hydro Seen Delayed," *The Financial Post*, 9 May 1942, 1; and "Quebec's Power Plans Delayed, Not Abandoned," 26 September 1942, 29.

19 The rumours of 1942 were aired and confirmed by *Le Soleil:* "Étatisation de l'électricité," 22 avril 1942, 1, 7; and "Le gouvernement étudie le projet d'étatisation, declare M. Godbout," 23 avril 1942. Regarding the plan to develop the Ottawa, the throne speech of 1943 stated that Godbout's regime would proceed to nationalize hydraulic powers of some 400,000 horsepower because Quebec had just signed an entente with Ontario to share the undeveloped waterpowers of the Ottawa River. The Legislative Assembly ratified the interprovincial agreement early in the session, but no project of nationalization followed.

20 These explanations are made in the newspaper articles noted above.

21 L.-P. Pigeon, "Mémoire confidential pour l'hon. M. Hamel," 2 février 1943, BANQQ, Fonds Godbout, dossier Hydro-électricité: Commission hydroélectrique de Québec.

22 "See Godbout's Plan As No Idle Threat," *The Financial Post*, 16 October 1943, news clipping in RN, Passe Dangereuse.

23 *Débats de l'Assemblée Législative*, séance du mardi 15 juin 1943, draft version. Said Chaloult, confusing the Peribonka and Saguenay projects:

"On avait promis de nationaliser l'électricité mais cette promesse ne semble pas devoir se réaliser. On cede un des plus grands pouvoirs hydrauliques de la province." The press conflated Shipshaw and Passe Dangereuse in "Des travaux de \$30,000,000 sur la rivière Péribonka," *L'Action Catholique*, 10 Septembre 1941 and "Travaux de plus de \$30,000,000 sur la rivière Péribonka," *L'Événement-Journal*, 11 Septembre 1941, news clippings in RN, Passe Dangereuse. Wrote Dubose to Latreille, 13 September 1941, of the articles' authors: "they were probably confusing the \$35,000,000 development with the Shipshaw extension [whereas] the dam on the Peribonka River would be a storage dam only ... to cost in the neighborhood of \$3,000,000."

24 On Alcoa's legal troubles, see Charles Carr, *Alcoa: An American Enterprise*, chaps. 15–17.

25 For example, "'Alum. of Canada' veut affermer les eaux du bassin du Saguenay," *L'Événement-Journal*, 26 septembre 1941; "Aluminum Ltd. of Canada," *Le Soleil*, 19 septembre 1941; and "L'enquête du Sénat américain sur l'aluminium," *L'Action Catholique*, 29 septembre 1941, clippings found in RN, Passe Dangereuse.

26 For example, "Coldwell accuse Ottawa de vendre le Saguenay à l'Aluminum Co.", *L'Événement-Journal*, 21 juin 1943, 10; "M. Coldwell accuse le gouvernement d'avoir favorisé les gros intérets," *Le Devoir*, 4 août 1943, 2.

27 That Godbout was pondering a general election is indicated in BANQQ, Fonds T.D. Bouchard, Box 15, file "Elections 1943," document "Discussions génerale au sujet des elections," 7 septembre 1943.

28 "Montreal Company Under Two-Way Fire," *The Financial Post*, 16 October 1943; "Provincial Power Plans Unsound," 9 November 1943, and "Power Grab Threat to Investors," *The Gazette* (Montreal), 11 November 1943.

29 "La Chronique Provinciale," 10 décembre 1943, clipping in BANQQ, Fonds Godbout, Électricité; "Hamel Visions Quebec 'Hydro,'" *The Gazette* (Montreal), 21 October 1943.

30 R. Rumilly, *Histoire de la Province de Québec*, vol. 40 (Montreal et Paris: Fides, 1969), 244–6. Rumilly also suggests a political motive in *Maurice Duplessis et son temps*, Tome 1 (Montreal: Fides, 1973), 654.

31 "La nationalisation de la Montreal Light, Heat and Power Consolidated," *L'Action Catholique*, 13 Octobre 1943, clipping in RN, Passe Dangereuse.

32 Hogue, et al., *Québec: un siècle d'électricité*, chap. 7; Claude Bellavance, Roger Levasseur, and Yvon Rousseau, "De la lutte antimonopoliste à la promotion de la grande enterprise, l'essor de deux institutions: Hydro-Québec et Desjardins, 1920–1965," *Recherches sociographiques* XL, 3,

1999: 551–78; Mario Dumais, "Étude sur l'histoire de l'industrie hydroélectrique (1940–1965) et son influence sur le développement industriel
au Québec" (Thèse de maîtrise, Université de Montréal, 1971).

33 Genest, *Godbout,* 263ff.

34 Gilles Gallichan, "De la Montreal, Light, Heat and Power à Hydro-
Québec," in *Hydro-Québec: Autres temps, autres defies,* Yves Bélanger and
Robert Comeau (Québec: Presses de l'Université du Québec, 1995),
64–70, 67: "C'est ce dernier rapport, remis au premier minister
en 1943, qui décide Godbout à agir"; Claude Bellavance, "Un long
mouvement d'appropriation de la première à la seconde nationalisation,"
Bélanger and Comeau, *Hydro-Québec,* 71–8, 74: "Il est certain que
ces renseignments ont joué un role déterminant dans la décision du
gouvernement Godbout d'étatiser la MLH & P"; and Robert Boyd,
"Cinquante ans au service du consommateur," Bélanger and Comeau,
Hydro-Québec, 97–103, 98: "si le gouvernement libéral d'Adélard Godbout,
en 1944, a crée Hydro-Québec et nationalisé le Montreal Light, Heat and
Power (MLH & P), c'est en raison de tarifs et de bénéfices jugés abusifs,
et d'un service jugé deficient."

35 Matthew Evenden, "La mobilisation des rivières et du fleuve pendant la
Seconde Guerre mondiale: Québec et l'hydroélectricité, 1939–1945,"
(a translation of the original manuscript: "Mobilizing Rivers: Global war,
Québec and Hydro, 1939–1945"), *Revue d'Histoire de l'Amérique Française,*
vol. 60, nos. 1–2 (été–automne 2006): 125–62.

CHAPTER 6

1 Massell, *Amassing Power.*

2 Evenden, "La mobilisation des rivières"; W.R. Way, "Power-System
Interconnection in Quebec," *Transactions of the American Institute of
Electrical Engineers* 61 (1942): 841–7; A.B. Normandin, "L'industrie hydroélectrique de Québec et l'effort de guerre," *Relations* novembre 1945,
298–301.

3 Minutes of Meeting of power officials, United States and Canada,
11 February 1942, Montreal, LAC, C.D. Howe Papers, Materials
Coordinating Committee.

4 Minutes of the Materials Coordinating Committee, 29 May 1942,
Montreal, LAC, C.D. Howe Papers, Materials Coordinating Committee.
Without unhindered access to the primary source holdings of Alcan
in Montreal, it is very difficult to determine executives' mindset in this
period of the war.

5 Dubose to Pierre Dupont, copy to R.E. Powell, 17 October 1942, file 00006 05, Rio Tinto Alcan, Montreal.

6 That a power shortage loomed was the impression of the business community generally in the spring of 1942. See, for example, *The Financial Post:* "Plan Cuts in Use of Electricity," 21 March 1942, 1; and "Power Shift May Be Made Late in Year," 18 April 1942, 1–2; and "Newsprint Looks at Aluminum," 10. As of mid-1942, Quebec's Public Service Board also forecast a power deficit of some 300,000 horsepower for the coming winter, according to Wilfrid Hamel ("Aucun contrôle sur l'électricité pour la fabrication du papier," *L'Événement-Journal,* 16 juin 1943, 1, 9).

7 Jacques Frenette, *Occupation et utilisation,* chap. 2.

8 Bird to Côté, 17 June 1942, Fonds Godbout, Aluminum Company of Canada; C.A. Robb to Dubose, 8 July 1942, found in BANQQ, Fonds Richesses naturelles, dossier Lacs Pipmaukin, des Prairies, Manouanis … Aluminum Co. of Canada, Ltd.

9 Dubose to Côté, 6 June 1942, and Dubose to Latreille (albeit missing page 2), 6 June 1942, BANQQ Fonds Godbout, Aluminum Company of Canada. A complete copy of both letters, with Latreille's handwritten notes, as well as much else of importance on this project, is found in BANQQ, Fonds Richesses naturelles, dossier Lacs Pipmaukin, des Prairies, Manouanis … Aluminum Co. of Canada, Ltd.

10 Latreille's notes in BANQQ, Richesses naturelles, Lac Pipmaukin.

11 "Extrait (minute 1060) du process-verbal de la C.E.C. du 23 juin 1942," BANQQ, Richesses naturelles, Lac Pipmaukin. Unfortunately the archives of the Quebec Streams Commission seem to have been lost or destroyed, as I have looked, in vain, for this material over several years. This meeting "minute" is but one type of historical document that these lost files once contained.

12 Latreille, "Aluminum Company … Résumé du projet," 10 juin 1942; Côté to Pigeon, 10 juin 1942; Côté to Dubose, 12 June 1942; Pigeon to Latreille, 26 juin 1942; all BANQQ Fonds Godbout, Aluminum Company of Canada, as well as BANQQ, Richesses naturelles, Lac Pipmaukin.

13 Geoffrion & Prud'homme to The Aluminum Company of Canada, 16 June 1942, BANQQ Fonds Godbout, Aluminum Company of Canada.

14 Bird, "Memorandum," 29 June 1942, BANQQ Fonds Godbout, Aluminum Company of Canada.

15 Powell to Godbout, 19 August 1942, BANQQ Fonds Godbout, Aluminum Company of Canada.

16 Avila Bédard to V.E. Bird, 21 August 1942, BANQQ, Richesses naturelles, Lac Pipmaukin.

17 Latreille (the likely author, although the memo may have been sent in the minister's name), "Mémoire confidential pour l'hon. M. Godbout," 7 juillet 1942, BANQQ, Richesses naturelles, Lac Pipmaukin.

18 Dubose to Latreille, 9 July 1942; Yvon DeGuise, "Résumé des projets de l'Aluminum Company Ltd," 23 juillet 1942; both BANQQ, Richesses naturelles, Lac Pipmaukin. The latter is also in BANQQ Fonds Godbout, Aluminum Company of Canada.

19 Latreille to Côté, 25 août 1942, BANQQ Fonds Godbout, Aluminum Company of Canada.

20 Latreille (likely) to Godbout, "Mémoire confidential pour l'hon. M. Godbout," 7 juillet 1942; Latreille to Côté, 25 août 1942; both BANQQ Fonds Godbout, Aluminum Company of Canada. Hudson's Bay Company to Latreille, 21 July 1942 and reply 27 July 1942, BANQQ, Richesses naturelles, Lac Pipmaukin.

21 Latreille to Côté, 30 août 1940, BANQQ Fonds Richesses naturelles, 1996-06-001/21, dossier Possibilité d'expansion des centrales hydroélectriques pour fins de guerre; Latreille to Côté, 28 mars 1942, BANQQ Fonds Godbout, Dossier Hydro-électricité.

22 Pigeon, "Mémoire confidential pour l'hon. Adélard Godbout La creation d'une hydro provinciale," 29 novembre 1943, BANQQ Fonds Godbout, dossier Électricité.

23 On Taschereau's politics, see Bernard Vigod, *Quebec Before Duplessis: The Political Career of Louis-Alexandre Taschereau* (Montreal & Kingston: McGill-Queen's University Press, 1986).

24 Pigeon, "Mémoire confidential pour l'hon. Adélard Godbout. Lois de l'électricité," handwritten, 20 mai 1940, BANQQ, Fonds Godbout, dossier Électricité. "A mon avis, il est souverainement important que vous ne donniez pas l'impression que vous rétablissez le régime d'avant 1936."

25 On the Toronto and Etobicoke plants, see Campbell, *Global Mission*, vol. 1, 162ff., 367–9.

26 Pigeon to Godbout, 3 août 1942; Pigeon informs Godbout of Aluminum's Ontario plant in Pigeon to Godbout, 19 juin 1942; BANQQ Fonds Godbout, Aluminum Company of Canada.

27 Pigeon to Godbout, 18 avril 1942; Latreille to Pigeon, 20 avril 1942; Latreille to Hamel, 12 mai 1943 (shared with Pigeon); BANQQ Fonds Godbout, Hydro-électricité.

28 Y. Deguise to A.B. Normandin, 27 juillet 1942, BANQQ, Richesses naturelles, Lac Pipmaukin.

29 Latreille, "Téléphone avec M.O. Lefebvre," 3 septembre 1942, BANQQ, Richesses naturelles, Lac Pipmaukin.

30 McCammon and Normandin to Côté, 4 septembre 1942; Lefebvre, Latreille, Normandin, and McCammon to Côté, 17 septembre 1942; and Latreille to Pigeon, 23 septembre 1942; BANQQ Fonds Godbout, Aluminum Company of Canada. Latreille to Lefebvre, and Latreille to Normandin, 24 août 1942; Côté to Lefebvre, and Côté to Normandin and McCammon, 25 août 1942; BANQQ, Richesses naturelles, Lac Pipmaukin.

31 Latreille to Côté, 25 août 1942, BANQQ Fonds Godbout, Aluminum Company of Canada. On Latreille's attempts to confirm the power shortage, see BANQQ, Richesses naturelles, Lac Pipmaukin, correspondence beginning 2 septembre 1942.

32 "Pas de disette d'électricité dans Québec," *Le Soleil*, 28 août 1942, 3.

33 "Une disette d'électricité, *Le Soleil*, 27 août 1942, 4.

34 "Disette d'électricité," *Le Progrès du Saguenay*, 8 octobre 1942, 1.

35 "Avant l'étatisation," *L'Événement-Journal*, 17 septembre 1942, 1.

36 "'Une politique pro-canadienne-française,' le mot d'ordre de l'avenir, dit M. Drouin," *L'Événement-Journal*, 7 septembre 1942, 3, 7. Drouin spoke before the annual youth congress of the provincial Fédération des Chambres de Commerce in La Sarre, Abitibi.

37 "M. Bouchard et l'étatisation de l'électricité," *L'Événement-Journal*, 17 septembre 1942, 3. Bouchard spoke at a dinner for Granby's Chamber of Commerce at the Hôtel Granby.

38 T.D. Bouchard, *Mémoires*, 157–8.

39 Fédération des Chambres de Commerce de la Province de Québec to Hon. A. Godbout, 13 janvier 1943, enclosing this and two other resolutions passed at the body's 7th Congress, 4–6 October 1942; and reply 21 janvier 1943; both BANQQ, Fonds Godbout, Box 5, Dossier 55, Électricité.

40 Latreille to Dubose, 29 July 1942, BANQQ, Richesses naturelles, Lac Pipmaukin.

41 Latreille to Normandin and McCammon, 14 novembre 1942, BANQQ Fonds Godbout, Aluminum Company of Canada.

42 Chs.-Ed. Deslauriers, "Notes Rivière aux Outardes," 7 octobre 1942; Deslauriers, "Rivière aux Outardes. Étude," 4 novembre 1942; A.W. Lash to Latreille, 15 October 1942; Latreille, "Note," 30 Novembre 1942; all BANQQ, Ressources naturelles, dossier Lac Pletipi, Tributaire rivière aux Outardes, Barrage reservoir.

43 Pierre Frenette, *Histoire de la Côte-Nord* (Québec: Institut québécois de recherche sur la culture, 1996), 367–72.

44 Arthur A. Schmon to Hon. Minister of Lands and Forests, 12 October 1942, BANQQ, Ressources naturelles, Lac Pletipi.

45 V. Edward Bird to W. Hamel, 3 May 1944, RN, Passe Dangereuse. Bird claimed that the government's denial of diversionary rights in 1942 resulted in "a very substantial power deficiency in the Saguenay [during the winter of 1943–44], necessitating unexpectedly heavy drafts upon the resources of other power companies in the province" as well as "unforeseen curtailments at our reduction plants in Arvida as well as at Shawinigan Falls, with a consequent disturbing effect upon labour."

46 Aluminum's interest in British Columbia is documented by Campbell, *Global Mission*, vol. 2, 51ff. It was also noted in "Aluminum Looks to Power Future," *The Financial Post*, 13 February 1943, 14.

CONCLUSION

1 Quinn, *The Union Nationale*, 81. Despite some recent re-examination of Duplessis's record and regime, Quinn's characterization remains intact. See Alain-G. Gagnon and Michel Sarra-Bournet, eds., *Duplessis: Entre la Grande Noirceur et la société libérale* (Montreal: Éditions Québec Amérique, 1997).

2 Roberts, *The Chief*, 3.

3 Campbell, *Global Mission*, vol. II, 39ff.

4 Genest, *Godbout*.

5 Quebec's perspective on the Quebec-Ontario accord can be gleaned from a reading of BANQQ, Fonds Godbout, dossier Accord de la Rivière Outaouais Quebec-Ontario. Duplessis is quoted by *L'Action Catholique*, 1 avril 1943, the exerpt found in this file. See also Evenden, "La mobilisation des rivières", 151–2.

6 Richard Rudolph and Scott Ridley, *Power Struggle: The Hundred-Year War over Electricity* (New York: Harper and Row, 1986), xi.

7 H.V. Nelles's "Public Ownership of Electrical Utilities in Manitoba and Ontario, 1906–30," *Canadian Historical Review* 57, 4 (1976): 461–84; Armstrong and Nelles, *Monopoly's Moment*, chap. 13. On Nova Scotia, see Lionel Bradley King, "The Electrification of Nova Scotia, 1884–1973: Technological Modernization as a Response to Regional Disparity" (Ph.D. dissertation, University of Toronto, 1999).

8 Nelles, "Public Ownership," 462.

9   Robert Blake Belfield, "The Niagara Frontier: The Evolution of Electric Power Systems in New York and Ontario, 1880–1935" (Ph.D. dissertation, University of Pennsylvania, 1981).

10 *Gaslights to gigawatts: A human history of BC Hydro and its predecessors*, BC Hydro Power Pioneers (Vancouver: Hurricane Press, 1998); David Schulze, "The Politics of Power: Rural Electrification in Alberta, 1920–1984" (master's thesis, McGill University, 1989).

11 Dales, *Hydroelectricity and Industrial Development*, 29–32.

12 Massell, *Amassing Power*, chaps. 4–5, and conclusion.

13 "Raymond Latreille Quitte La Commission Hydroélectrique de Québec," *Hydro-Presse*, 1966, Archives d'Hydro-Québec.

14 Godbout to J.A. Habel, 1 avril 1947, BANQQ, Fonds Godbout, Accord de la Rivière Outaouais Quebec-Ontario.

15 Canadian Indians won federal voting rights, without qualification related to property or sex, in 1960. The list of provinces granting the franchise to Indians begins with British Columbia (1949), then Manitoba (1952), Ontario (1954), Saskatchewan (1960), Prince Edward Island and New Brunswick (1963), and Alberta (1965). Quebec (1969) came last in this chronology. The subject is summarized in the *Report of the Royal Commission on Aboriginal Peoples* (Ottawa: 1996), vol. 1, chap. 9, §9.12.

16 Conférence de presse de M. Raymond Latreille 14 avril 1964 à l'occasion de 20ième anniversaire de l'Hydro-Québec, 4, Fonds Jean Lésage, 1986-03-007/43, file 24, BANQQ.

17 See Paul Charest, "Les barrages hydro-électriques en territoire Montagnais", as well as the transcriptions of interviews with Native hunters which were conducted by Gilbert Courtois in 1977–78, on which Charest's article was based: *Étude sur l'impact des barrages hydroélectriques*. Charest has also summarized his findings, in English, in "Hydroelectric dam construction and the foraging activities of eastern Quebec Montagnais," in *Politics and history in band societies*, eds. Eleanor Leacock and Richard Lee (Cambridge: Cambridge University Press, 1982), 413–26.

18 On Cree co-operation with Hudson's Bay Company agents in the creation and implementation of the James Bay beaver preserves, see Hans Carlson, *Home is the Hunter: The James Bay Cree and Their Land* (Vancouver: University of British Columbia Press, 2008), chap. 5; and Tina Loo, *States of Nature: Conserving Canada's Wildlife in the Twentieth Century* (Vancouver: University of British Columbia Press, 2006), 102ff. The Peribonka counterpoint is revealed in LAC, Department of Indian Affairs (RG 10), vols. 420-10-4-1, 420-10-4, 420-10-1-1, 420-10-1-3. Several factors seem

to have hampered the effectiveness of the scheme from the beginning, including Montagnais skepticism of the project, greater opportunities for Native wage labour in an increasingly industrializing region, and greater competition and conflict among Indians themselves over family trapping grounds (leading to Native poaching) due to a half-century's encroachment by white settlers and lumbermen along the Saguenay to the south. "Native attitude is not as favourable as in some preserves," project supervisor Hugh Conn admitted in his Peribonka Beaver and Fur Preserve, Fifth Annual Report, 1945; and in the 1947 Report: "the native attitude leaves much to be desired."

19 Interview with Gérard Siméon, Informateur 159, regarding the migration of 1945, 10, MAM: "Avant qu'il y ait un barrage, ils ne pouvaient monter les rapides Kapitatshutsh [Passe Dangereuse]. C'est après l'inondation que les 'Blancs' ont commencé à monter."

20 P. Charest, "Les barrages hydro-électriques en territoire Montagnais."

21 Bernier to Dubose, 15 February 1952, and Cousineau to Dubose, 13 February 1953, in file 00014 22, Alcan, Montreal. Both Bernier and Cousineau were responding to Dubose's requests for information regarding the effects of the Saguenay's dams and storage reservoirs on fish and game. Dubose was engaged in the vast Kitimat-Kemano project of British Columbia and was contending with controversy over the project's potential damage to salmon populations.

22 "Le carcajou brisait tout: les caches, les tentes et il pissait sur les provisions ... Aujourn'hui, c'est pareil, le Blanc c'est comme le carcajou." Fiches d'entrevue de l'informateur 150 (Joseph Bellefleur), CIP, interview conducted by Mathieu Paul, 1981.

23 "Quand l'homme [blanc] touche à quelque chose, il le détruit. Il n'y que la pierre que l'homme blanc ne peut détruire." Entrevue de périodisation, Informateur 13, 8, MAM.

24 P. Charest, "Les barrages hydro-électriques en territoire Montagnais."

25 See, for example, Ken Coates, *Best Left as Indians: Native-White Relations in the Yukon Territory, 1840–1973* (Montreal & Kingston: McGill-Queen's University Press, 1991), 189; Kerry Abel, *Drum Songs: Glimpses of Dene History*, 2nd ed. (Montreal & Kingston: McGill-Queen's University Press, 2005, 1st ed. 1993), chap. 9; or Ken Coates and William Morrison, *The Alaska Highway in World II: The U.S. Army of Occupation in Canada's Northwest* (Norman: University of Oklahoma Press, 1992), chap. 3.

26 P. Charest, "Les barrages hydro-électriques en territoire Montagnais"; also D. Brassard, *Occupation et utilisation*, and J. Frenette, *Occupation et utilisation*.

27 Clark, "Rivers of Aluminum," 226, 231. That Clark's manuscript was deemed inappropriate for publication by Alcan's Public Relations staff is understandable, given its paternalistic, at times condescending, portrayal of French Canadians and Natives. Where Lac Manouan's dam construction is concerned, for example, Clark recounts an anecdote (p. 229) of company pilots forced by winter weather to make an emergency landing at a wilderness food depot along the supply route, a depot that was supposed to be overseen by two Montagnais. Neither the Indians nor the food was found, however, and for three days the crew subsisted on cigarettes and a snared rabbit before the weather cleared. "Subsequent investigation," Clark writes, "revealed the Bucks had been making love to a couple of young squaws in a camp across the lake, using the food as bait."

28 "Le Lac Péribonka, çela a fait tort ... il n'y a plus rien ... il y en avait du rat musqués ... puis aujourd'hui pas un. Il y en a pas, ils ont rien à manger. C'est tout baigné ça aujourd'hui, il n'y a que bois pourri." Témoignage de Jack Germain in Courtois, *Étude sur l'impact des barrages hydroélectriques.*

29 Hudson's Bay Company Archives, Pointe-Bleue (B. 329) and Bersimis (B. 17), Post Journals.

30 Frank Tough, *'As Their Natural Resources Fail': Native Peoples and the Economic History of Northern Manitoba, 1870–1930* (Vancouver: University of British Columbia Press, 1996), 301–2.

31 "Dans la rivière elle-même, tu es pas capable de vivre là. Par rapport qu'il y a trop d'inconvenients, il y a l'eau qui monte et qui baisse, tu sais, ça joue de meme tout le temps, tu peux pas tendre de piéges la, c'est foutu ... Oui, j'ai travaillé à la construction de ce barrage-là aux Passes. Mais dans ce temps-là par exemple, tu sais, nous autres on prévoyaient pas, tu sais, que s'ils envoyaient un barrage de meme ... Si on auraient prévu que cela aurait pu faire du tort, endommagé ... qu'on détruisaient nos biens ... [mais] personne parlait, ni ne disait un mot." Témoignage de Gérard Siméon in Courtois, *Étude sur l'impact des barrages hydroélectriques.*

32 Conversation with Gérard Siméon, June 2006; author's translation. Denis Brassard catalogs the social upheavals of the post-war period in *Occupation et utilisation*, chaps. 4–5.

33 Paul Charest, "Les ressources naturelles de la Côte-Nord ou la richesse des autres: une analyse diachronique," *Recherches amérindiennes au Québec* 5, 2 (1975): 35–52.

34 José Igartua, "'Corporate' Strategy and Locational Decision-Making:
    The Duke-Price Alcoa Merger, 1925," *Journal of Canadian Studies* 20, 3
    (1985-86): 82-101.

35 E.G. MacDowell to R.E. Powell, January 1944, quoted by Campbell, *Global
    Mission*, vol. II, 7.

36 P. Whitney Lackenbauer, *Battle Grounds: The Canadian Military and
    Aboriginal Lands* (Vancouver: University of British Columbia Press, 2007).
    On a related topic, see Sarah Carter, "'An Infamous Proposal': Prairie
    Indian Reserve Land and Soldier Settlement After World War I," *Manitoba
    History* 37 (1999): 9-21. For the larger historiographical context, see
    Lackenbauer and R. Scott Sheffield, "Moving Beyond 'Forgotten': The
    Historiography on Canadian Native Peoples and the World Wars," in
    Aboriginal Peoples and the Canadian Military: Historical Perspectives, eds.
    P. Whitney Lackenbauer and Chris Leslie Mantle (Kingston: Canadian
    Defence Academy Press, 2007), 209-31.

37 Morris Zaslow, *The Northward Expansion of Canada, 1914-1967* (Toronto:
    McClelland and Stewart, 1988), 209.

38 Conversation with Gérard Siméon, June 2006.

39 James Waldram, *As Long as the Rivers Run: Hydroelectric Development and
    Native Communities in Western Canada* (Winnipeg: University of Manitoba
    Press, 1988).

40 Camil Girard, "Reconnaissance historique des peuples autochtones
    au Canada: Territories et autonomie gouvernementale chez les Innu
    (Montagnais) du nord-est du Québec de 1603 à nos jours," unpublished
    paper, 2004, shared by the author; Douglas Sanders, *Native People in Areas
    of Internal National Expansion: Indians and Inuit in Canada* (Copenhagan:
    International Workgroup for Indigenous Affairs, no. 14, 1973).

41 See, for example, Paul Charest, "The Land Claims Negotiations of the
    Montagnais, or Innu, of the Province of Quebec and the Management
    of Natural Resources," in *Aboriginal Autonomy and Development in Northern
    Quebec and Labrador,* ed. Colin H. Scott (Vancouver: University of British
    Columbia Press, 2001), 255-73; and Bernard Cleary, "Le long et difficile
    portage d'une negotiation territoriale," *Recherches amérindienne au Québec*
    23, 1 (1993): 49-60.

42 On this subject, see Claude Bellavance and Pierre Lanthier, *Les Territoires
    de l'entreprise, The Territories of Business* (Sainte-Foy: Les Presses de
    l'Université Laval, 2004).

43 Bothwell and Kilbourn, *C.D. Howe*, p. 274.

44 Howe Papers, vol. 170, Howe-Powell correspondence of June 1944, LAC.

45 Historian Daryl White, in a careful study of wartime relations between the Canadian government and both Alcan and Inco, has judged that "the spirit of Alcan's cooperation with the Canadian authorities in the Second World War was rarely in doubt," nor was the corporation's "commitment to Allied victory." Unfortunately, White also did not have access to internal corporate documents in order to better test executives' motives in this regard. See White, "Multinational Patriots: Business-Government Relations in the Canadian Aluminum and Nickel Industries, 1914–1945" (Ph.D. dissertation, University of Western Ontario, 2006), 314, 316.

46 Crerar, "History of the Development of the Saguenay Power System"; Clark, "Rivers of Aluminum"; Campbell, *Global Mission*, vol. II.

47 Hydro-Québec had already tinkered with the Peribonka watershed in 2003 by diverting the upper Manouan River (at an enlarged Lac du Grand Détour) into the Bersimis's Lac Pipmaukan, thereby adding minor additional flow volume (some 30 cubic meters per second) to its Bersimis hydroelectric complex.

48 Paul Charest reviews recent so-called 'partnership'agreements between Hydro-Québec and the Innu in "More Dams for Nitassinan: New Business Partnerships Between Hydro-Québec and Innu Communities," in *Power Struggles*, eds. Thibault Martin and Steven M. Hoffman, 255–79.

49 Morris Zaslow, *The Opening of the Canadian North, 1870–1914* and *The Northward Expansion of Canada, 1914–1967* (Toronto: McClelland and Stewart, 1971 and 1988).

50 On the environmental history of war, see Edmund Russell, *War and Nature: Fighting Humans and Insects with Chemicals from World War I to Silent Spring* (Cambridge: Cambridge University Press, 2001); and Richard Tucker and Edmund Russell, eds., *Natural Enemy, Natural Ally: Toward an Environmental History of Warfare* (Corvallis: Oregon State University Press, 2004). Particularly applicable to our study here is Richard Tucker's essay "The World Wars and the Globalization of Timber Cutting." Writes Russell in his study of chemical warfare and pest control (p. 2): "the control of nature expanded the scale of war, and war expanded the scale on which people controlled nature."

51 Massell, *Amassing Power*, chaps. 4–5.

52 Tucker, in his essay "The World Wars and the Globalization of Timber Cutting," makes clear that the world wars also increased state intervention in the management of forest industries.

Bibliography

PRIMARY SOURCES

Manuscripts

Alcan Aluminium Limited (Rio Tinto Alcan), Business Information
   Centre, Montreal:
   The Aluminum Ingot
   Corporate memoirs (various)
   Photographs
Archives d'Hydro-Québec
   Hydro-Presse
   Photographs
   Fonds Régie Provinciale de l'électricité
Bibliothèque et Archives nationales du Québec, Montreal:
   Fonds Régie Provinciale de l'électricité
Bibliothèque et Archives nationales du Québec, Québec:
   Cour Supérieure, Québec
   Fonds Adélard Godbout
   Fonds Jean Lésage
   Fonds Conseil exécutif Québec
   Fonds Ressources naturelles
   Fonds Richesses naturelles
   Fonds Télesphore Damien Bouchard
Centre d'expertise hydrique du Québec, Quebec:
   Aluminum of Canada Ltd.
Conseil des Innus de Pessamit, Pessamit/Betsiamites, Quebec:
   Entrevues de périodisation
   Fiches de description du site de campement
   Fiches d'entrevue

Franklin D. Roosevelt Presidential Library, Hyde Park, New York:
    Adolf Berle Papers
    Herbert Marks Papers
    Morgenthau Diaries
    Leland Olds Papers
Hudson's Bay Company Archives:
    Bersimis (B. 17)
    Pointe-Bleue (B. 329)
Musée amérindien de Mashteuiatsh, Mashteuiatsh/Pointe-Bleue, Quebec:
    Entrevues de périodisation
    Fiches de description du site de campement
    Fiches d'entrevue
Ministère des Ressources naturelles et de la Faune, Charlesbourg, QC, office
    of développement électrique:
    Dossier Lac Manouan
    Dossier Passe Dangereuse/Rivière Péribonca
    Dossier Rivière Saguenay (Grande Décharge)
Library and Archives Canada:
    MG 26 J, William Lyon Mackenzie King
    MG 27 III B 20, C.D. Howe
    MG 30–D45 John Wesley Dafoe
    RG 2, Privy Council Office
    RG 10, Department of Indian Affairs
    RG 14, Parliament
    RG 19, Department of Finance
    RG 28, Department of Munitions and Supply
National Archives and Records Administration of the United States, College
    Park, Maryland:
    Reconstruction Finance Corporation
    War Production Board
St-Joseph d'Alma, QC: Léopold Naud Private Collection

Government documents

Annual Reports of the Department of Indian Affairs (Canada)
Débats de l'Assemblée Législative (Quebec)
Debates, House of Commons (Canada)
Foreign Relations of the United States, Washington: Department of State
Hearings Before a Special Committee to Investigate the National Defense
    Program [the Truman Committee], United State Senate

Hearings Before the Special Committee to Study and Survey Problems
of American Small Business Enterprises (Problems of American Small
Business), United States Senate
Hearings Before the Subcommittee of the Committee on Appropriations,
United States Senate on S.J. Res. 285, 10 July 1940
Quebec Sessional Papers
Rapports de la Commission des eaux courantes de Québec
Reports of the Minister of Lands and Forests (Quebec)
Statutes of the Province of Quebec

Newspapers
L'Action Catholique
Le Devoir
L'Événement-Journal
The Financial Post
The Gazette (Montreal)
The New York Times
La Presse
Le Progrès du Saguenay
Le Soleil

Interviews
Alain Nepton, June 2006
Gérard Siméon, June 2006
Adélard Bellefleur, May 2008
Marthe Godbout-Bussières, June 2009

SECONDARY SOURCES

Abel, Kerry. *Drum Songs: Glimpses of Dene History*, 2nd ed. Montreal & Kingston:
    McGill-Queen's University Press, 2005 (1st ed., 1993).
– "Tangled, Lost and Bitter? Current Directions in the Writing of Native
    History in Canada," *Acadiensis* 26, 1 (1996): 92–101.
Amos, Arthur. *Les Forces Hydrauliques de la Province de Québec*. Québec:
    Département des Terres et Forêts, 1917.
Anderson, J.W. *Fur Trader's Story*. Toronto: Ryerson Press, 1961.
Armstrong, Christopher, and H.V. Nelles. "Contrasting Development of
    the Hydro-Electric Industry in the Montreal and Toronto Regions,
    1900–1930." *Journal of Canadian Studies* 18, 1 (spring 1983): 5–27.

– *Monopoly's Moment: The Organization and Regulation of Canadian Utilities,*
   1830–1930. Philadelphia: Temple University Press, 1986.
Barrette, Antonio. *Mémoires.* Montreal: Librairie Beauchemin, Limitée, 1966.
Beaulieu, Alain. "Du nomadisme aux reserves; histoire et culture des
   Montagnais du Québec." In *Les Indiens Montagnais du Québec: Entre deux*
   mondes, edited by Anne Vitart, 11–33. Paris et Québec: Éditions Sépia,
   1995.
Beaulieu, Carl. *B.A. Scott: Père de l'industrialisation.* Chicoutimi, QC: Les
   Éditions Entreprises, 1999.
Bédard, Hélène. *Les Montagnais et la Réserve de Betsiamites, 1850–1900.* Québec:
   Institut québécois de recherche sur la culture, 1988.
Belfield, Robert Blake. "The Niagara Frontier: The Evolution of Electric
   Power Systems in New York and Ontario, 1880–1935." Ph.D. dissertation,
   University of Pennsylvania, 1981.
Bellavance, Claude. *Shawinigan Water and Power 1898–1963: formation et déclin*
   d'un groupe industriel au Québec. Montreal: Les Éditions du Boréal, 1994.
– "L'état, la 'houille blanche' et le grand capital: L'aliénation des ressources
   hydrauliques du domaine public québécois au début du XXe siècle," *Revue*
   d'histoire de l'Amérique française 51, 4 (1998): 487–520.
– and Roger Levasseur, Yvon Rousseau. "De la lutte antimonopoliste à la
   promotion de la grande enterprise, l'essor de deux institutions: Hydro-
   Québec et Desjardins, 1920–1965." *Recherches sociographiques,* XL, 3, 1999:
   551–78.
– "Un long mouvement d'appropriation de la première à la seconde
   nationalization." In *Hydro-Québec: Autres temps, autres defies,* edited by Yves
   Bélanger and Robert Comeau, 71–8. Québec: Presses de l'Université du
   Québec, 1995.
– and Pierre Lanthier. *Les Territoires de l'entreprise, The Territories of Business.*
   Sainte-Foy: Les Presses de l'Université Laval, 2004.
Blanchard, Raoul. *L'Est du Canada Français,* Tome 2. Paris: Librairie Masson,
   1935.
Blough, Earl, et al. "Discussion on the Design of the Shipshaw Power
   Development." *The Engineering Journal* 27, 8 (August 1944): 449–58.
Boismenu, Gérard. *Le Duplessisme: Politique économique et rapports de force,*
   1944–1960. Montreal: Les Presses de l'Université de Montréal, 1981.
Bolduc, André, and Clarence Hogue, Daniel Larouche. *Québec: un siècle*
   d'électricité. Montréal: Éditions Libre Expression, 1979.
Bothwell, Robert, and William Kilbourn. *C.D. Howe: A Biography.* Toronto:
   McClelland and Stewart Limited, 1980.

Bouchard, Russel. "De Saint-Amédée à Chute-des-Passes: la colonisation de la Péribonca," *Saguenayensia* 37, 3/4 (1995): 3–23.

Bouchard, T.D. *Mémoires*. Montreal: Éditions Beauchemin, 1960.

Boudreault, René. *Mashteuiatsh*. Wendake: Institut culturel et éducatif Montagnais, 1994.

Boyd, Robert. "Cinquante ans au service du consommateur." In *Hydro-Québec: Autres temps, autres defies*, edited by Yves Bélanger and Robert Comeau, 97–103. Québec: Presses de l'Université du Québec, 1995.

Brassard, Denis. *Occupation et utilisation du territoire par les Montagnais de Pointe-Bleue*. Village-des-Hurons: Conseil Attikamek-Montagnais, 1983.

Brock, T.L. "Alcan in the Saguenay – the Formative Years." Montreal: Aluminium Company of Canada, Ltd, 1971.

Bryce, Robert B. *Canada and the Cost of World War II: The International Operations of Canada's Department of Finance, 1939–1947*, edited by Matthew J. Bellamy. Montreal & Kingston: McGill-Queen's University Press, 2005.

Burgesse, J. Allan. "Property Concepts of the Lac-St-Jean Montagnais." *Primitive Man* 43 (1945): 1–25.

Campbell, Duncan C. *Global Mission: The Story of Alcan*. 3 volumes. Montreal: Alcan Aluminium Limited, 1985, 1990.

Carlson, Hans. *Home Is the Hunter: The James Bay Cree and Their Land*. Vancouver: University of British Columbia Press, 2008.

Carlson, Keith Thor, Melinda Marie Jetté, and Kenichi Matsui. "An Annotated Bibliography of Major Writings in Aboriginal History, 1990–99." *Canadian Historical Review* 82, 1 (2001): 122–71.

Carr, Charles. *Alcoa: An American Enterprise*. New York: Rinehart and Company, Inc., 1952.

Chaloult, René. *Mémoires politiques*. Montreal: Éditions du jour, 1969.

Charest, Paul. "Les barrages hydro-électriques en territoire Montagnais et leurs effets sur les communautés amérindiennes." *Recherches amérindiennes au Québec* IX, 4 (1980): 323–37.

– "Chevauchements de territoires entre la bande de Betsiamites et les bandes voisine." Betsiamites: Conseil de bande de Betsiamites, Bureau politique, 2001.

– "Hydroelectric dam construction and the foraging activities of eastern Quebec Montagnais." In *Politics and history in band societies,* edited by Eleanor Leacock and Richard Lee, 413–26. Cambridge: Cambridge University Press, 1982.

– "The Land Claims Negotiations of the Montagnais, or Innu, of the Province of Quebec and the Management of Natural Resources." In *Aboriginal Autonomy and Development in Northern Quebec and Labrador,* edited by Colin H. Scott, 255–73. Vancouver: University of British Columbia Press, 2001.

– "Les ressources naturelles de la Côte-Nord ou la richesse des autres: une analyse diachronique." *Recherches amérindiennes au Québec* 5, 2 (1975): 35–52.

"Consultation avec les membres des familles et des trappeurs ayant fréquenté la region nord-est du territoire de la réserve à castor de Bersimis." Betsiamites: Conseil de bande de Betsiamites, Bureau Politique, Secteur negotiation, 1998.

Clark, Paul. "Rivers of Aluminum: The Story of Alcan." 1978 manuscript held by Alcan in Montreal.

Cleary, Bernard. "Le long et difficile portage d'une negotiation territoriale." *Recherches amérindiennes au Québec* 23, 1 (1993): 49–60.

Coates, Ken. *Best Left as Indians: Native-White Relations in the Yukon Territory, 1840–1973.* Montreal & Kingston: McGill-Queen's University Press, 1991.

– "Writing First Nations into Canadian History: A Review of Recent Scholarly Works." *Canadian Historical Review* 81, 1 (2000): 99–114.

Comeau, Napoleon A. *Life and Sport on the North Shore of the Lower St Lawrence and Gulf.* Quebec: Daily Telegraph Printing House, 1909.

Côté, Dany. *De la terre, du bois, de l'eau et des gens: de Honfleur à Sainte-Monique, 1898–1998.* Sainte-Monique: Municipalité de Sainte-Monique, 1997.

– *Histoire de l'industrie forestière du Saguenay-Lac-Saint-Jean.* Publication no. 17, Société d'histoire du Lac-Saint-Jean, 1999.

– *Riverbend: Splendeur et déclin d'une ville de compagnie.* Publication no. 8, Société d'histoire du Lac-Saint-Jean, 1994.

Courtois, Gilbert. *Étude sur l'impact des barrages hydroélectrique.* Unpublished, MAM, 1978.

Crerar, N.S. "History of the development of the Saguenay Power System." Transcript of address given before the Electrical Club of Montreal, 2 April 1958. Copy available at Alcan Aluminium Limited, Montreal.

Cuff, R.D., and J.L Granatstein. *Ties That Bind: Canadian-American Relations in Wartime from the Great War to the Cold War,* 2nd ed.. Toronto and Sarasota: Samuel Stevens, Hakkert and Company, 1977.

Dales, John. *Hydroelectricity and Industrial Development 1898–1940.* Cambridge: Harvard University Press, 1957.

Danforth, R.T. *Aluminum Company of America and Subsidiaries, Massena, New York, Chronological Notes.* Copy obtained at Alcoa Primary Metals, North-East Smelting Region, Massena Operations.

Deschênes, Jean-Guy, and Richard Dominique. *Nitasinan.* Village-des-Hurons: Conseil Attikamek-Montagnais, 1983.

Dubose, McNeely. "The Engineering History of Shipshaw." *The Engineering Journal* 27, 4 (April 1944): 194–9.

Dumais, Mario. "Étude sur l'histoire de l'industrie hydro-électrique (1940– 1965) et son influence sur le développement industriel au Québec." Thèse de maîtrise, Université de Montréal, 1971.

Electric Power and Government Policy: A Survey of the Relations Between the Government and Electric Power Industry. New York: The Twentieth Century Fund, 1948.

Evenden, Matthew. *Fish Versus Power: An Environmental History of the Fraser River.* Cambridge: Cambridge University Press, 2004.

– "La mobilisation des rivières et du fleuve pendant la Seconde Guerre mondiale: Québec et l'hydroélectricité, 1939–1945." *Revue d'Histoire de l'Amérique Française,* 60, 1–2 (été–automne 2006): 125–62.

– "Lights Out: Conserving Electricity for War in the Canadian City, 1939– 1945." *Urban History Review* 34, 1 (2005): 88–99.

Fleury, Jean Louis. *Une Ligne et des homes.* Montréal: Libre Expression, 1991.

Francis, Daniel. *A History of the Native Peoples of Quebec, 1760–1867.* Ottawa: Minister of Indian Affairs and Northern Development, 1983.

Fraser, Blair. "Victory in Aluminium." *Maclean's Magazine,* 1 February 1944, 13ff.

Frenette, Jacques. *Occupation et utilisation du territoire par les Montagnais de Betsiamites, 1920–1982.* Village-des-Hurons: Conseil Attikamek-Montagnais, 1983.

Frenette, Pierre. *Histoire de la Côte-Nord.* Québec: Institut québécois de recherche sur la culture, 1996.

– and Dorothée Picard. *Histoire et culture innues de Betsiamites.* Betsiamites: École Uashkaikan/Les Presses du Nord, 2002.

Froschauer, Karl. *White Gold: Hydroelectric Power in Canada.* Vancouver: University of British Columbia Press, 1999.

Funigiello, Philip J. *Toward a National Power Policy: The New Deal and the Electric Utility Industry, 1933–1941.* Pittsburgh: University of Pittsburgh Press, 1973.

Gagnon, Alain-G., and Michel Sarra-Bournet, eds., *Duplessis: Entre la Grande Noirceur et la société libérale.* Montreal: Éditions Québec Amérique, 1997.

Gallichan, Gilles. "De la Montreal, Light, Heat and Power à Hydro-Québec." In *Hydro-Québec: Autres temps, autres defies,* edited by Yves Bélanger and Robert Comeau, 64–70. Québec: Presses de l'Université du Québec 1995.

Gaslights to gigawatts: A human history of BC Hydro and its predecessors. BC Hydro Power Pioneers, Vancouver: Hurricane Press, 1998.

Genest, Jean-Guy. *Godbout.* Sillery: Les editions du Septentrion, 1996.

– "L'Élection provinciale québécoise de 1939." Thèse (histoire), Université Laval, 1968.

Gill, Pierre. *Les Montagnais, premiers habitants du Saguenay-Lac-St-Jean.* Pointe-Bleue: Mishinikan, 1987.

Girard, Camil, and Normand Perron, *Histoire du Saguenay-Lac-St-Jean.* Quebec: Institut québécois de recherche sur la culture, 1995 (1989).

– "Reconnaissance historique des peoples autochtones au Canada: Territories et autonomie gouvernementale chez les Innu (Montagnais) du nord-est du Québec de 1603 à nos jours." Unpublished paper, 2004.

Girard, Jeannette, and Jean-François Moreau. "Histoire et préhistoire de la rivière Péribonca." *Saguenayensia* 29, 1 (1987): 6–12.

Granatstein, J.L. *Canada's War: The Politics of the Mackenzie King Government, 1939 1945.* Toronto: Oxford University Press, 1975.

Guttman, Frank Myron. *The Devil from Saint-Hyacinthe: Senator Télesphore-Damien Bouchard.* Lincoln, NE: iUniverse, 2007.

Ickes, Harold. *The Secret Diary of Harold L. Ickes,* vol. III. New York: Simon and Shuster, 1954.

Igartua, José E. *Arvida au Saguenay: naissance d'une ville industrielle.* Montreal & Kingston: McGill-Queen's University Press, 1996.

– "'Corporate' Strategy and Locational Decision-Making: The Duke-Price Alcoa Merger, 1925." *Journal of Canadian Studies* 20 (1985–86): 82–101.

Industrial Mobilization for War: History of the War Production Board and Predecessor Agencies, 1940–1945, vol. 1. New York: Greenwood Press, 1969.

James, Robert Warren. *Wartime Economic Co-operation.* Toronto: Ryerson Press, 1949.

Janeway, Eliot. *The Struggle for Survival: A Chronicle of Economic Mobilization in World War II.* New Haven: Yale University Press, 1951.

Jauvin, Serge. *Aitnanu: The Lives of Hélène and William-Mathieu Mark,* edited by Daniel Clément. Montreal: Libre Expression, 1993.

Jeffries, Hugh. "Up north, there's mystery." *Liberty,* 26 April 1941.

Kennedy, J. de N. *History of the Department of Munitions and Supply: Canada in the Second World War.* Canada: Edmond Cloutier/King's Printer, 1950.

Kerwin, Jerome G. *Federal Water-Power Legislation.* New York: Columbia
  University Press, 1926.
King, Lionel Bradley. "The Electrification of Nova Scotia, 1884–1973:
  Technological Modernization as a Response to Regional Disparity." Ph.D.
  dissertation, University of Toronto, 1999.
Lacasse, Jean-Paul. *Les Innus et le territoire.* Sillery: Éditions du Septentrion,
  2004.
Lackenbauer, P. Whitney. *Battle Grounds: The Canadian Military and Aboriginal
  Lands.* Vancouver: University of British Columbia Press, 2007.
– and Craig Leslie Mantle, eds. *Aboriginal Peoples and the Canadian Military:
  Historical Perspectives.* Kingston: Canadian Defence Academy Press, 2007.
Lawton, F.L. "The Manouan and Passe Dangereuse Water Storage
  Developments." *The Engineering Journal* 27, 4 (April 1944): 200–20.
Leacock, Eleanor. *The Montagnais "Hunting Territory" and the Fur Trade.* Memoir
  No. 78, *Memoirs of the American Anthropological Association,* 1954.
Lemoine, Géo. *Dictionnaire Français-Montagnais avec un vocabulaire Montagnais-
  Anglais, une courte liste de noms géographiques et une grammaire Montagnaise.*
  Boston: W.B. Cabot and P. Cabot, 1901.
Lips, Julius. *Naskapi Law (Lake St John and Lake Mistassini Bands): Law and Order
  in a Hunting Society. Transactions of the American Philosophical Society,* vol. 37,
  pt. 4 (December 1947): 379–492.
– "Notes on Montagnais-Naskapi economy (Lake St John and Lake Mistassini
  Bands)." *Ethnos,* vol. 12, nos. 1–2 (January–June 1947): 1–78.
Litvak, Isaiah A., and Christopher J. Maule. *Alcan Aluminium Limited, A Case
  Study.* Royal Commission on Corporate Concentration Study No. 13,
  Ottawa: Minister of Supply and Services Canada, 1977.
Loo, Tina. *States of Nature: Conserving Canada's Wildlife in the Twentieth Century.*
  Vancouver: University of British Columbia Press, 2006.
Mabee, Carleton. *The Seaway Story.* New York: The Macmillan Company, 1961.
Marlio, Louis. *The Aluminum Cartel.* Washington: The Brookings Institution,
  1947.
Martin, Jean. "Le Triomphe Canadien de la Deuxième Guerre Mondiale:
  L'Émergence d'Alcan dans l'Industrie Nordaméricaine de l'Aluminium."
  Unpublished paper, summer 2004.
Martin, Thibault and Steven M. Hoffman, eds. *Power Struggles: Hydro
  Development and First Nations in Manitoba and Quebec.* Winnipeg: University
  of Manitoba Press, 2008.
Massell, David. *Amassing Power: J.B. Duke and the Saguenay River, 1897–1927.*
  Montreal & Kingston: McGill-Queen's University Press, 2000.

– "'As though there was no boundary': the Shipshaw Project and Continental Integration." *The American Review of Canadian Studies* (Summer 2004): 187–222.

– "Power and the Peribonka, a Prehistory: 1900–1930s." *Quebec Studies*, vol. 38, Fall 2004/Winter 2005, 87–103.

McGuire, B.J. and H.E. Freeman. "How the Saguenay River Serves Canada." *The Canadian Geographical Journal* 35, 5 (1947): 200–25.

McIntyre, Paul-Emile. "The Development of Hydro-electric Power at Grand Falls, New Brunswick, an issue in provincial politics, 1920–1926." Master's thesis, University of New Brunswick, 1973.

Miller, John T. *Foreign Trade in Gas and Electricity in North America: A Legal and Historical Study*. New York: Praeger Publishers, 1970.

Morantz, Toby. *The White Man's Gonna Getcha: The Colonial Challenge to the Crees of Quebec*. Montreal & Kingston: McGill-Queen's University Press, 2002.

Nappi, Carmine. "Canada: An Expanding Industry." *The World Aluminum Industry in a Changing Energy Era*, edited by Merton J. Peck, 175–221. Washington: Resources for the Future, 1988.

Nash, Gerald D. *World War II and the West: Reshaping the Economy*. Lincoln: University of Nebraska Press, 1990.

Nelles, H.V. "Public Ownership of Electrical Utilities in Manitoba and Ontario, 1906–30." *Canadian Historical Review* 57, 4 (1976): 461–84.

Nelson, Donald M. *Arsenal of Democracy: The Story of American War Production*. New York: Harcourt, Brace and Company, 1946.

Nobbs, Percy E. *Salmon Tactics*. Boston and New York: Houghton Mifflin Company, 1935.

Normandin, A.B. "L'industrie hydro-électrique de Québec et l'effort de guerre" *Relations* Novembre 1945, 298–301.

Oliver, Michael. *The Passionate Debate: The Social and Political Ideas of Quebec Nationalism, 1920–1945*. Montreal: Véhicule Press, 1991.

Owen, Marguerite. *The Tennessee Valley Authority*. New York: Praeger Publishers, 1973.

Pickersgill, J.W. *The Mackenzie King Record*, vol. 1. Chicago and Toronto: University of Chicago Press, University of Toronto Press, 1960.

Pouyez, C., Y. Lavoie, et al. *Les Saguenayens: Introduction à l'histoire des populations du Saguenay XVIe–XXe siècles*. Sillery: Presses de l'Université du Québec, 1983.

Quinn, Herbert. *The Union Nationale: A Study in Quebec Nationalism*. Toronto: University of Toronto Press, 1963.

Ray, Arthur. *The Canadian Fur Trade in the Industrial Age*. Toronto: University of Toronto Press, 1990.

Report of the Royal Commission on Aboriginal Peoples, vol. 1. Ottawa: 1996.

Roberts, Leslie. *The Chief: A Political Biography of Maurice Duplessis.* Toronto and Vancouver: Clarke, Irwin and Co. Ltd., 1963.

Rudolph, Richard, and Scott Ridley. *Power Struggle: The Hundred-Year War over Electricity.* New York: Harper and Row, 1986.

Rumilly, Robert. *Histoire de la Province de Québec,* Tomes 36–41. Montreal et Paris: Fides, 1966–69.

– *Maurice Duplessis et son temps,* Tome I (1890–1944). Montreal: Fides, 1973.

Russell, Edmund. *War and Nature: Fighting Humans and Insects with Chemicals from World War I to Silent Spring.* Cambridge: Cambridge University Press, 2001.

Sanders, Douglas. *Native People in Areas of Internal National Expansion: Indians and Inuit in Canada.* Copenhagan: International Workgroup for Indigenous Affairs, no. 14, 1973.

Schulze, David. "The Politics of Power: Rural Electrification in Alberta, 1920–1984." Master's thesis, McGill University, 1989.

Scott, James. *Seeing Like a State.* New Haven and London: Yale University Press, 1998.

Simard, Robert. "La mission d'Onistagan," *Saguenayensia* 13, 2 (mars–avril 1971): 51–54; and 13, 3 (mai–juin 1971): 69–73.

Siméon, Anne-Marie, and Camil Girard. *Un monde autour de moi: Témoignage d'une Montagnaise.* Chicoutimi: Les éditions JCL, 1997.

Sitterson, J. Carlyle. *Aircraft Production Policies Under the National Defense Advisory Commission and Office of Production Management May 1940 to December 1941.* Historical Reports on War Administration: War Production Board, Special Study No. 21, 1946.

Smith, Philip. *It Seems Like Only Yesterday: Air Canada, the First 50 Years.* Toronto: McClelland and Stewart, 1986.

Speck, Frank. "Family Hunting Territories of the Lake St John Montagnais and Neighboring Bands." *Anthropos* 22 (1927): 387–403.

Tough, Frank. *'As Their Natural Resources Fail': Native Peoples and the Economic History of Northern Manitoba, 1870–1930.* Vancouver: University of British Columbia Press, 1996.

Tremblay, Victor. *La Tragédie du lac St-Jean.* Chicoutimi, QC: Editions Science Moderne, 1979.

– "La rivière Péribonka: les premières pages de son histoire," *Saguenayensia* 1, 6 (nov–déc 1959): 143–6.

– "La rivière Péribonka: la période des explorations," *Saguenayensia* (jan–fév 1960): 17–24.

– "La rivière Péribonka: période des chantiers et de la colonization, *Saguenayensia* (sept–oct 1973): 134–62.

Tucker, Richard, and Edmund Russell, eds. *Natural Enemy, Natural Ally: Toward an Environmental History of Warfare.* Corvallis: Oregon State University Press, 2004.

Veilleux, André. "Pointe-Bleue: Histoire d'une reduction." Master's thesis (sociology), Université Laval, 1982.

Vigod, Bernard. *Quebec Before Duplessis: The Political Career of Louis-Alexandre Taschereau.* Montreal & Kingston: McGill-Queen's University Press, 1986.

Waldram, James. *As Long as the Rivers Run: Hydroelectric Development and Native Communities in Western Canada.* Winnipeg: University of Manitoba Press, 1988.

Wallace, Donald. *Market Control in the Aluminum Industry.* Cambridge: Harvard University Press, 1937.

Way, W.R. "Power-System Interconnection in Quebec." *Transactions of the American Institute of Electrical Engineers* 61 (1942): 841–7.

Wiltse, Charles. *Aluminum Policies of the War Production Board and Predecessor Agencies May 1940 to November 1945.* Historical Reports on War Administration: War Production Board, Special Study No. 22, 1946.

Whitaker, Albert. *Aluminum Trail.* Montreal: Alcan Press, 1974.

White, Daryl F. "Multinational Patriots: Business-Government Relations in the Canadian Aluminum and Nickel Industries, 1914–1945." Ph.D. dissertation, University of Western Ontario, 2006.

Worster, Donald. Nature's Economy: A History of Ecological Ideas. Cambridge: Cambridge University Press, 2nd ed., 1994.

Zaslow, Morris. *The Opening of the Canadian North, 1870–1914.* Toronto: McClelland and Stewart, 1971.

– *The Northward Expansion of Canada, 1914–1967.* Toronto: McClelland and Stewart, 1988.

Index

Abenaki, 24, 38

Abitibi region, 29

Action catholique, 140

Action libérale nationale, 149

Advisory Commission to the Council of National Defense (United States), 101–2, 105. *See also* Office of Production Management; War Production Board

aerial survey, 29–30, 32, 35, 144

aerial warfare, 100, 106

agriculture, 15, 83, 158, 167. *See also* tragédie du lac Saint-Jean

Air Canada, 105

Alberta, 157

Alcoa Power Company, 54

Aluminium Limited, 14, 17, 53

aluminum, 5, 102–3, 106–12. *See also* Aluminum Company of America; Aluminum Company of Canada; Second World War

Aluminum Company of America (Alcoa), 5, 13–14, 17, 52–4, 96, 101–3, 111–13, 133, 138, 150–1, 155, 170

Aluminum Company of Canada (Alcan): background on, 5, 11, 13–14; expansion in Saguenay

watershed, 6, 9, 17, 52–64, 77–98, 107–17, 143–56, 159, 167–71; and Great Depression, 32; investigation/criticism of, 133–6; maps/illus., 68–76. *See also* Rio Tinto

Amos, Arthur, 19, 27, 31, 60, 158–9. *See also* Hydraulic Service

Anderson, J.W., 22, 24–5

André Daniel, 36, 38

animism, 46, 50. *See also* Innu

Arvida, 53–4, 106, 115, 133, 139, 161; map, 65

Ashuapmushuan River, 17, 28, 153; maps, xiv–xv

Attikamek, 11, 35

Bacon family, 23, 39, 46

Baie-Comeau, 152–3

Baie des Hirondelles, 44–6

Bank of Canada, 109

Batt, William, 101, 106, 196n14

Battle of Britain, 59, 106

bauxite, 59, 106

Beauharnois dam, 103

Beauharnois Light, Heat and Power, 137–8

beaver preserves, 161, 214n18

Bédard, Avila, 56, 60, 85, 92

Bédard, Hélène, 24

Bégin family, 50

Bélanger, H., 30

Bellefleur, Adélard, and family, 43, 45–7

Bellefleur, Joseph, 162

Benjamin, Paul, and family, 23, 43–5

Berle, Adolf Jr., 103

Bernier, Omer, 161

Bersimis (Betsiamites/Pessamit) Reserve, 4, 11–13, 20, 23–4, 37, 144, 161–2, 164, 167; illus., 127; maps, xiv–xv. *See also* Innu

Bersimis River (Rivière Betsiamites), 20; Aboriginal travel on, 37, 40, 44–7; concessions sought on, 144–54, 169, 171; maps, xiv–xv. *See also* Pessamit

Betsiamites. *See* Bersimis Reserve

Bibliothèque et Archives nationales du Québec, 11

Big Otter River (Rivière de la Grande Loutre), 41, 44–5; map, xiv

Bignell, John, 16

Bird, V. Edward, 85, 92

Bloc populaire canadien, 136

Blough, Earl, 108

Boivin, Baso (Joseph), and family, 35–9, 43, 45

Boivin, Charles, 35–6

Boivin, Louis-Georges, and family, 3–6, 35

Bonneville dam, 101

Bonneville Power Administration, 157

Bouchard, Télesphore-Damien, 139–40, 151–2, 155; illus., 124

Bourget River, 42

Bourque, John, 57

Brassard, Denis, 51

British Columbia, 54, 153

British Columbia Power Commission, 157

British Guyana, 59

Broadback River, 29

Brodeuse River (Rivière Brodeuse), 20, 39–42, 44; map, xiv

Brown, Frank, 108, 111

Brown Corporation, 147

Bucknell, Charlot, 43

Burgesse, J. Allan, 50

Cabinet (Quebec), 80–1, 85, 91–4, 146, 151, 168

Canada: shortage of US dollars, 109–11; US power imports from, 101–17

Canadian Geographical Journal, 8

Canadian Geological Survey, 16

Canadian National Railways, 105

Canadian Shield, 8, 165, 170

Carillon, 156

Casey, Joseph, 42

Catawba River, 16

Cedar Rapids Manufacturing and Power Company, 102

Chaloult, René, 135–6, 155

Chamberlain, Neville, 5

Charest, Paul, 165

Charlotte, NC, 16

Chicago Tribune, 152

Chicoutimi, 42, 64, 83, 136, 151

Christianity/Catholicism, 24, 46–7, 50

Churchill Falls project, 8

Churchill River (Manitoba), 166

Chute-à-Caron, 32, 35, 54, 57, 79–80, 82, 93, 103, 106–7, 148

Chute-à-la-Savane, 29, 57, 169; illus.,

70; map, xv

Chute-du-Diable, 29, 57, 169; illus., 70, map, xv

Chutes-des-Passes, 169; map, xv

Chutes McLeod, 36, 38, 40, 43–4

Clark, Clifford, 110–11, 114

Clark, Paul, 77, 95, 163, 216n27

Clayton, W.L., 111

Coffee, John, 133

Coldwell, M.J., 133–4, 139

Collard, Malec: illus., 131

Collège Mont-Saint-Louis, 158

Columbia River, 90, 101, 157

Comeau, N.A., 25

Commission des eaux courantes. See Quebec Streams Commission

Conn, Hugh, 214n18

Connolly, Elie, and family, 43–4, 50

Connolly, Emery, 38

conscription, 60, 136

Consolidated Edison Company, 113

Consolidated Paper Corporation, 84–7

Co-operative Commonwealth Federation, 133, 139

Cooter, David Eaton, 50

Côté, Georges, 30

Côté, Pierre-Émile, 60, 80–1, 83, 85, 91–2, 95–6, 115–17, 137, 144–5, 150–1. See also Ministry of Lands and Forests (Quebec)

Cothran, Frank, 55, 187n43

Courtois, Gilbert, 163

Cousineau, Yvon, 161–2

Cran Serré, 41, 43–4,

Cree, James Bay, 25, 45, 51, 161. See also Mistassini, Lake

Czechoslovakia, 5

Dafoe, John W., 104

Davis, Arthur Vining, 52–3

Davis, Edward, 53–4, 108

Defense Plant Corporation, 113, 203n63

Deguise, Yvon, 94

Department of Finance (Canada), 109–11, 114

Department of Indian Affairs (Canada), 24, 51

Department of Lands and Forests (Quebec). See Ministry of Lands and Forests (Quebec)

Department of Munitions and Supply (Canada), 105, 108, 111

Department of the Interior (United States), 101

Depression. See Great Depression

Desbiens, Joseph and William, 38

Desterres, Joseph: illus., 131

Dolbeau, 44, 94

Dominique, Jean-Baptiste, and family, 40, 42–3, 162

Dominique, Joseph: illus., 126

Dominique, Michel, and family, 35, 38–40

Dominique, Paul, 45

Doucet, Pierre, 49

Drolet, Joseph 'Tit-Rouge,' 37, 43, 49

Drouin, Oscar, 135, 151–2, 155

Dubose, McNeely, 54, 56, 60–1, 78–81, 86, 90–3, 107, 143–5, 147, 153, 161, 168; illus., 121. See also Bersimis River, concessions sought on; Lac Manouan, concession of; Lac Passe Dangereuse, concession of

Du Chef River, 17

Dufour, Clément, 37, 43

Dufresne Engineering Company, 94

Duguay, Joseph-Léonard, 56

Duke, James Buchanan, 16–17, 27–9, 52–3, 95, 142; illus., 119
Duke-Price Power Company, 29–31, 55, 97
Dunn, Gano, 101, 103, 196n14
Duplessis, Maurice, 9, 154–6, 169; and critique of Godbout's concessions to Alcan, 63, 96, 134–36, 138–9, 141; and Lac Manouan concession, 56–9; illus., 122
Du Pont Company, 170; illus., 119
Duval, Éloi, 28

Eastmain River, 41; map, xiv
Eastman Kodak Company, 103
École Polytechnique de Montréal, 158
Edison Electric Institute, 101
election of 1935 (Quebec), 149
election of 1936 (Quebec), 58
election of 1939 (Quebec), 60, 149, 155
electrical power. See public power; Second World War
Electrical World, 100
Electricity Commission. See Quebec Electricity Commission
electrochemical manufacture, 90. See also Aluminum Company of America; Aluminum Company of Canada
emphyteutic lease, 11, 80
Étienne, Patrick and family, 21–2
Evenden, Matthew, 141

farmers. See agriculture
Fawley, Mr, 50
Federal Power Commission (United States), 99–102, 104

Fédération des chambres de commerce, 152
Financial Post, The, 137, 139
fire wardens, 49
First World War, 170
fishing/fisheries, 49, 64, 87–90, 135, 159, 161
Fontaine family, 23
forestry. See timber
forks, the (Peribonka-Manouan confluence): and Aboriginal travel, 3, 21, 37–8, 40–4; dam built at, 169; dam planned nearby, 17, 31
fluorspar, 59
Frankland, H.B., 163
Franklin D. Roosevelt Presidential Library, 11
Fraser River, 90
Frenette, Jacques, 51
fur trade, 3, 48–50, 52, 159–60, 164. See also Innu

Gagnon, Adolphe, 24
Gagnon, Joseph, 48–9
Gatineau Power, 11
Gatineau River, 83
Gazette, The, 139
Geoffrion, Aimé, 60, 64, 80, 92, 95–7, 114, 146, 189n64
Germain, Barthélemy and François, and families, 36, 38–40, 45, 163
Germain, Jack, and family, 20–1, 43, 50, 55, 163–4; illus., 132
Germain, Michel, 40
Germain, Noé, 45
Germany, 5–6, 54
Girardville, 49
Giroux, Louis, 29
Glenn L. Martin Company, 110

Godbout, Adélard, 9, 59, 89, 114, 154–6; and concession of Lac Manouan, 59–64; and concession of Lac Passe Dangereuse, 78–98, 115–17; and concessions sought on North Shore rivers, 144–53; Opposition attacks on and creation of Hydro-Québec, 134–41; papers of, 11, 95, 97, 138; illus., 122–3

Gouin, Lomer, 158

Gouin (La Loutre) Reservoir, 28, 55, 79, 145, 149

Grand Coulee dam, 101

Grand Discharge, 148; map, 65

Grand Lac Détour, 44

Grasse River, 102

Great Britain, 5–6, 54, 59, 61, 103–4, 106, 109, 114, 165

Great Depression, 32–3, 35, 49, 51, 56, 107, 138, 152, 156–7

Gustafson, G.H., 84

Hall, Charles Martin, 13

Hamel, Wilfrid, 89, 92–3, 134–5, 137–8, 153

Hébertville, 64

Hémon, Louis, 8, 15

Hirondelles River, 46

Hitler, Adolf, 5

Holden, Grenville, 103

Honfleur, 28, 36, 38; maps, xiv–xv, 66

Howe, C.D., 105–11, 113–15, 168; illus., 118

Hudson River, 110

Hudson Strait, 19

Hudson's Bay Company, 20, 24, 42, 48–51, 147, 160, 163–4

Hyde Park Declaration, 110–11, 116

Hydraulic Service (Quebec): history of/background on, 10–11, 18–19, 27, 33, 82, 84. See also Amos, Arthur; Lac Manouan, concession of; Lac Passe Dangereuse, concession of; Latreille, Raymond; Normandin, A.B.

hydroelectricity: historiography of, 176–8n16–18; and Native people, 159, 162–3; Quebec policy regarding, 10–11, 33–4, 55, 60, 86–7, 90, 96, 154–9, 169–70; source of cheap power, 6, 150; subsidiary to pulp/paper production, 19, 26; technical development of, 16, 142. See also Hydro-Quebec; Lac Manouan; Lac Passe Dangereuse; nationalization of electricity

Hydro-Québec, 7, 10, 18, 83, 94, 117, 149, 137–41, 155 8. 169–71, 175n14; illus., 124–5. See also nationalization of electricity

Ickes, Harold, 133

Igartua, José, 165

Innu (Montagnais), 3, 6, 11, 19–25, 35–52, 159–67, 180n17; 126–32; maps, xiv–xv

Isle Maligne dam, 16, 29–30, 32, 34–5, 38, 53, 61, 79–80, 82, 88, 93, 106–7, 142; maps, 65, 69

James Bay, 29, 161

James Bay project, 8–9, 160, 171, 176n15

Japan, 5, 54, 97

Joncas, Paul, 26

Jourdain family, 50

Kak'wa family, 50

Kapu, Sylvestre, and family: illus., 128–9

Kellogg, Charles, 101–4

Killam, Isaak Walton, 104

King, Mackenzie, 109–10, 115–16, 133, 136

Kingston, 133

Kitimat-Kemano project, 153, 215n21

Knudsen, William, 106

Krug, Julius, 99, 101, 196n14

Kurtness family, 50

Lac à Michel, 40

Lac à la Montagne, 21

Lac Benoit, 39

Lac des Prairies, 144

Lac Manouan (Manouan Lake): Aboriginal travel on, 42, 44; concession of, 8–10, 55–64, 82, 96, 149, 151, 154–5, 169; dam to impound, 64, 88, 95; government survey of, 32; Native participation in dam construction, 163–4; Opposition criticism of, 134–6, 138; scale of, 79, 191n9; illus. 67, 69–70, 71–6

Lac Manouanis, 47, 144

Lac Maurice, 46

Lac Onistagane, 25, 36, 39–40, 42; maps, xiv–xv

Lac Opitoonis, 32; map, 67

Lac Napissi, 41

Lac Natipi, 41

Lac Passe Dangereuse (now called "Lac Péribonka"): concession of, 8–10, 55, 77–98, 114, 149, 151, 154–5, 169; dam construction for, 94–5; funding for, 111; Native participation in survey and construction of, 163–5; Opposition criticism of, 134–6, 138, 141; scale of, 8, 79, 81–2, 145–6; and Shipshaw dam, 112, 115, 138; illus., 74–6; maps, xv, 69–70

Lac Péribonka (note: this is the current name of Passe Dangereuse reservoir, as well as the name of the largest of the chain of lakes that was flooded to form Lac Manouan reservoir), 42; map, 67. See also Lac Manouan, Lac Passe Dangereuse

Lac Pipmuacan, 44–7; as reservoir, 46, 146–7; maps, xiv–xv

Lac Pletipi, 48, 146, 152–3; maps, xiv–xv

Lac St-Jean (Lake St John): Aboriginal travel on/around, 3, 20, 22, 38, 41; concession of, 52, 61, 80, 82; flood control for, 30–2, 83, 135; impoundment as storage reservoir, 4, 7, 17, 27–30, 35, 55, 95, 106, 142–3, 148; insufficient to regulate the Saguenay, 17–18, 55, 153; illus., 69–70; maps, xiv–xv, 65. See also Isle Maligne dam; tragédie du lac Saint-Jean

Lac Tchitogama, 42

Lac Trippe d'Ours, 40

Lackenbauer, P. Whitney, 166

Langelier, J.C., 19

Lapointe, Erneste, 116

Latreille, Raymond, 31, 63–4, 78–82, 84, 87, 89–94, 96, 117, 137, 144–8, 150–3, 155, 158–9, 167, 169; illus., 124–5

Lawton, Fred, 55, 144, 163

Leclerc, Georges, 26

Lee, William States, 16–17, 34, 142; illus., 119

Lefebvre, Olivier, 82–3, 92, 96, 150–1, 155

Legislative Assembly (Quebec), 11, 78–9, 81, 93, 145, 149

Lend-Lease Act, 109

Le Progrès du Saguenay, 151

Le Soleil, 151

L'Heureux, Eugene, 140

Liberal Party (Canada), 104, 116

Liberal Party (Quebec), 58, 117, 134–7, 149–50. *See also* Godbout, Adélard; Taschereau, Louis Alexandre

Library and Archives Canada, 11

Lièvre River, 83

Lips, Julius, 49–51

Long Sault, 102, 104

Low, A.P., 16

MacDowell, Elmer, 54, 59, 81, 111

MacLaren-Quebec Power, 11

Manicouagan River: Aboriginal travel on, 4, 20, 37, 40; concessions sought on, 147–53, 171; 125, 131–2, 152–3; reservoirs of, 8; surveys of, 29; maps, xiv–xv

Manitoba, 104, 156, 164, 166

Manitoba Free Press, 104

Manouan Dam. *See* Lac Manouan

Manouan River: Aboriginal travel on, 3–4, 43–6; relation to Bersimis River, 144; surveys of, 26, 30, 32, 88; illus., 69, 70; maps, xiv–xv

Martin, Adhémar: illus., 131

Mashteuiatsh. *See* Pointe-Bleue Reserve

Massena, NY, 101–4, 112–13, 116;
map, 68

Mathewson, James Arthur, 114

Mattawin River, 83

McCammon, John, 117, 143, 150–1, 200n34; illus., 124

McCormick, Robert, 152

McGill University, 87

McTague, C.P., 114–15

Mégiscane River, 145

Mellon, Andrew and Richard, 13

Metals Reserve Company, 77, 81, 111–13, 115–16

Miller, Charles, 163

Miller, Joseph, 45, 48–9

mining, 52, 83–4, 115, 158, 160

Ministry of Fish and Game (Quebec), 87–9

Ministry of Lands and Forests (Quebec), 10–11, 19, 26–7; Surveys Branch of the, 29, 35. *See also* Côté, P.E.; Hamel, Wilfrid; Lac Manouan, concession of; Lac Passe Dangereuse, concession of

Ministry of Mines (Quebec), 84

Ministry of Natural Resources and Wildlife (Quebec), 10

missionaries. *See* Christianity/ Catholicism

Mistassibi River, 20, 44; maps, xiv–xv, 65, 69

Mistassini, Lake, 16, 153; Native band of, 22, 25, 41, 44, 50; maps, xiv–xv

Mistassini River, 20, 28; maps, xiv–xv, 65, 69

Moar, Mr, 40

Moar family, 50

Modeste River, 44; map, xiv

Montagnais. *See* Innu

Montreal, 90, 104, 137, 152, 156,

158. *See also* Aluminum Company of Canada; Consolidated Paper Corporation; Quebec Pulp and Paper Corporation; Quebec Streams Commission

Montreal Light, Heat and Power, 117, 137–8, 156, 158; illus., 124

Morgenthau, Henry Jr, 110–11

Musée amérindien de Mashteuiatsh, 11

Nappi, Carmine, 9

Naskapi, 50

National Archives and Records Administration of the United States, 11

National Defense Advisory Commission. *See* Advisory Commission to the Council of National Defense

National Defense Power Committee/ National Power Policy and Defense Committee, 100

nationalization of electricity, 7, 59, 134–41, 151, 155–8, 171

Natipi, Joseph, and family, 23, 36, 41–3

Native people. *See* Innu

Nelson River, 166

Nepton, Philippe and Willie, 164

Netaukat, 48

New Brunswick, 156

New Deal, 157

Newfoundland, 59

new staples, 52

New York, 90, 102, 104, 112–13, 157

New York Power Authority, 157

New York Times, The, 54, 133

Niagara Falls, 103, 156

Nipissi River, 44

Nobbs, Percy Erskine, 87–90

Normandin, A.B., 56–8, 60–2, 64, 95, 150–1, 154, 158, 188n52

North Carolina/Carolinas, 54, 102

North Shore region/Côte Nord, 152, 154, 165, 171; maps, xiv–xv. *See also* Bersimis River; Manicouagan River; Outardes River; Quebec North Shore Paper Company

Northrup Aircraft Incorporated, 110

Nottaway River, 29

Nova Scotia, 156

Obedjiwan Reserve, 35

Oberlin College, 13

Office of Production Management (United States), 106

Ogdensburg, NY, 116

Olds, Leland, 101–4

Ontario, 33, 59, 102–4, 138, 148, 150, 156–7

Ontario Hydro-Electric Commission, 105, 156–7

Ottawa, 6–7, 51, 89, 97, 102, 104, 107, 115–16, 136–7, 152, 168, 170

Ottawa River, 29, 138, 150, 156

ouananiche. *See* fishing/fisheries

Outardes River (Rivière aux Outardes): Aboriginal use of, 20, 37; concessions sought on, 146–53, 171; survey of, 29; maps, xiv–xv

Pacific Northwest, 102; map, 68

Parc des Laurentides, 23

Paré, A. Euclide, 94

Parliament (Canada), 113, 139

Pashushipashtuk, 43

Passe Dangereuse, 36, 39–42, 44, 78,

87. *See also* Lac Passe Dangereuse

Pearl Harbor, 97

Péribonka IV, 169; map, xv

Peribonka River (Rivière Péribonka),
8; Aboriginal travel on, 3–4, 20–5,
35–51; hydro concessions of, 6–7,
154–9, 169; industrial history of,
15–18; surveys of, 26, 28–31, 142;
maps, xiv–xv, 65–6, 69–70. *See also*
Bersimis River; Lac Manouan; Lac
Passe Dangereuse

Péribonka (village), 15

Pessamit, 45, 185n17. *See also*
Bersimis Reserve

Petawabano family, 22

Philippe, David, and brother, 38–41,
49

Piakunuk, 45

Picard, Charles Henri, and family, 47

Picard family, 23, 43, 46

Piekukemiuluts, 45

Pigeon, Louis-Philippe, 93, 138, 145,
148–51, 155

Pipmuacan River, 47, 152

Pittsburgh Reduction Company, 13,
101

Plekushi, 46

Plowden, E.C., 165

Pointe-Bleue Reserve (Mashteuiatsh),
3–4, 12–13, 161, 163–4, 167, 170,
186n31; journeys from/to, 20–5,
35–44, 47–50; illus., 126, 130;
maps, xiv–xv. *See also* Innu

Poitras, Felix: illus., 131

Poland, 5

Powell, Ray E., 32, 54, 59, 107–9,
111–12, 114, 134, 168; and
concession of Lac Manouan,
61–2, 96; and concession of Lac
Passe Dangereuse, 78–81, 92, 97;

and concession of North Shore
rivers, 143, 146; illus., 120

Power Controller (Canada), 105,
116, 135, 141, 143. *See also*
Symington, Herbert

Price, William, and Price Brothers,
16, 26, 56, 94

Progressive Era/progressivism, 18,
33–4, 59–60, 82, 107, 140, 156–7.
See also Godbout, Adélard

Provencher, Paul, 51; photographs
by, 128–9, 131–2

Prud'homme, J. Alexandre, 79, 85,
91

public power, 101, 156–7

Public Service Board (Régie des
services publics), 64, 95, 105, 115,
117, 138, 140, 143, 150–1, 156,
199n34

pulp and paper, 16, 19, 26, 56, 104,
135, 149, 169. *See also* timber/
logging/sawmills

Quebec City, 6, 51, 60, 81, 115, 142,
152, 160

Quebec Development Company, 95,
97. *See also* Duke, James Buchanan

Quebec Electricity Commission, 137,
140, 149

Quebec North Shore Paper
Company, 147, 152–3

Quebec Province: territorial growth
of, 19. *See also* hydroelectricity

Quebec Pulp and Paper Corporation,
84–7, 192n27

Quebec Streams Commission, 27–8,
31, 33, 35, 56, 82–3, 92, 96,
145–6, 150–1, 169, 210n11

Raphael family, 43; illus., 130

Rapide Trenche, 147–8

Ray, Arthur, 48

Reconstruction Finance Corporation (United States), 81. *See also* Metals Reserve Company

Richard, L.A., 87–90

Rio Tinto, 9, 11, 13–14

Rivière à la Carpe, 37

Rivière à Michel, 37, 39; map, xiv

Rivière aux Outardes. *See* Outardes River

Rivière aux Sables, 44

Rivière Betsiamites. *See* Bersimis River

Rivière Manicouagan. *See* Manicouagan River

Roberval, 16, 36, 38, 64; maps, xiv–xv, 65

Rock family, 23

Roosevelt, Franklin D., 53, 99–101, 103, 110, 133, 157

Roy, Léon, 36, 49

Royal Securities, 104

Rumilly, Robert, 139

Rupert River, 153; maps, xiv–xv

Saguenay-Lac St-Jean region: agricultural settlement of, 15, 23; economy of, 90

Saguenay Power Company, 55–6, 60, 78, 163–4. *See also* Dubose, McNeely

Saguenay River: flow regulation of, 4–5, 16–17; power development of, 8, 16, 52–3, 106, 110, 142–4; sale of bed and powers of, 82; illus., 70; maps xiv–xv, 65, 69–70. *See also* Aluminum Company of America; Aluminum Company of Canada; Duke, James Buchanan;

Lac Manouan; Lac Passe Dangereuse

St-Ambroise, 42

St-Félicien, 37, 49; maps, xiv–xv, 65

St-Henri-de-Taillon, 15

St-Hyacinthe, 152

St-Jean, 136

St Lawrence aluminum plant, 113

St Lawrence River, 101–2, 112, 143–4, 150, 153; illus., 127; maps, xiv–xv, 65, 69

St Lawrence Seaway, 103–4, 197n20

St-Maurice River, 28, 33, 35, 79, 143, 147, 149

St-Onge, Paul, and family, 36, 43, 46

St-Onge River, 44; map, xiv

Ste-Marguerite River, 168; map, xv

Ste-Monique-de-Honfleur, 15. *See also* Honfleur

Sapin Croche River, 20

Sarnia, 104

Saskatchewan River, 166

Savane River, 37, 39–42; map, xiv

Savard, François, 40

Schmon, Arthur, 153

Scott, Benjamin, 16–18, 26, 142; illus., 119

Second World War, 5; and aluminum production, 9–10, 32–3, 35, 54, 58–64, 85, 100–1, 106–9, 143, 165–6, 168–71; and electrical production, 99–104; and nationalization of electricity, 137, 141, 156–7; and Native people, 162–3

Serpent River, 36, 39–42, 44; map xiv

Shawinigan Falls, 13

Shawinigan Water and Power Company, 11, 143, 147–9

Sherbrooke, 152

Shipshaw dam, 17, 54, 77–9, 81–2, 92–3, 103, 106–15, 133, 138, 142–3, 148, 168; maps xv, 68–70
Shipshaw River, 16, 53, 147; maps, xiv–xv
Shipshaw Scandal, 133, 155
Siméon, Anne-Marie, 22, 25
Siméon, Gérard, and family, 43, 161, 164–6
Siméon, Malek, and family, 22, 36, 40, 43
Siméon, Thomas, and family, 43–4
Simpson, Willie, 164
Smith, Philip, 104
Southern Power Company, 16
Speck, Frank, 24
Stettinius, Edward, 101, 106, 196n14
Strauss, S.D., 111, 113
Sugar River (Rivière au Sucre), 42
Surveys Branch. *See* Ministry of Lands and Forests (Quebec)
Sutherland, James, 19
Sylvestre, family, 38, 40
Sylvestre River (Rivière Sylvestre), 46
Symington, Herbert, 77, 104–12, 115–17, 134, 137, 143, 145, 168; illus., 118

Taschereau, Louis Alexandre, 58–9, 95, 135, 149, 158
Témiscamie River, 41; maps, xiv–xv
Tennessee Valley Authority, 82, 101, 157
timber/logging/sawmills, 16, 26, 35, 49, 52; and hydro concessions, 64, 78–9, 84–7, 134–6, 145, 149, 154, 158, 160–1, 168; illus., 76. *See also* pulp and paper
Toronto, 150
Tough, Frank, 164

Toupin, T., 28
tourism, 24, 38, 49, 87–90
tragédie du lac Saint-Jean, 4, 30, 35, 55, 83, 95
Trans-Canada Airlines, 105; illus., 118
Tremblay, Aldège, 37
Tremblay, William, 26
Trinidad, 59
Triton Fish and Game Club, 38
Truman, Harry, 110
Tsernish family, 23, 45
Tsheketash family, 23

Ungava Peninsula, 19
Union Nationale, 58, 135–6, 140. *See also* Duplessis, Maurice
United States, 6–7, 54, 77, 81, 97–117, 148, 155–7, 165; Congress of, 101, 104, 109. *See also* Aluminum Company of America; Reconstruction Finance Corporation; Metals Reserve Company
Université du Québec à Chicoutimi, 170
University of Toronto, 104
urbanization, 48, 142

Vachon, Madame Joseph: illus. 132
Valin, William, 22
Vladykov, Vadim D., 88

Waldram, James, 166
Wall Street Journal, 99
War Committee/War Cabinet (Canada), 109–10, 113
War Contracts Depreciation Board (Canada), 114
War Department (United States), 100

War Exchange Conservation Act
  (Canada), 114
War Expenditures Committee
  (Canada), 115, 133–4
War Measures Act (Canada), 115
War Production Board (United
  States), 99, 106
Washington, D.C., 6–7, 81, 102,
  105, 111, 115–16. *See also* United
  States
Washish family, 23
Water-Course Act, 94–5
Wheat Boom, 104

White Mountain River (Rivière
  Montagne Blanche), 3, 5, 42;
  map, xiv
Willson, Thomas, 19, 142
Wilmington, 170
Winnipeg, 104
workers/labourers, 94–5, 159, 164,
  174n4; illus., 71–2
World War Two. *See* Second World War

Xavier, territory of, 36

Zaslow, Morris, 166, 170